D0102389

# THE UNFOLDING UNIVERSE

PATRICK MOORE

# THE UNFOLDING UNIVERSE

MICHAEL JOSEPH/RAINBIRD

To Pieter Morpurgo, without whom
this book would not have been written

First published in Great Britain in 1982
by Michael Joseph Ltd
44 Bedford Square, London WC1 and
The Rainbird Publishing Group Ltd
40 Park Street, London W1Y 4DE
who designed and produced the book



ISBN 07181 2152 X

Edited by Caroline Zelkha
Designed by Martin Bristow
Illustrated by Paul Doherty

Text set by SX Composing Ltd, Rayleigh, Essex, England
Colour originated by Lithospeed Ltd, London, England
Printed and bound by New Interlitho SPA, Milan, Italy

*Endpapers:* A distant cluster of galaxies. Many of the objects shown here are not stars in our own Galaxy, but external systems containing many thousands of millions of stars. Foreground stars can also be seen. Photograph taken with the Anglo-Australian Telescope (AAT), Siding Spring, New South Wales

*Half-title page:* Eclipse of the Sun by Neptune, as seen from Triton. The Sun appears only as a brilliant star; refraction in Neptune's atmosphere gives the appearance of a complete ring. Painting by Paul Doherty

*Frontispiece:* The nebula NGC 6164/5. More properly this is the nebulosity around a very hot star, HD 148937, which is the brightest member of a system made up of three stars in orbit round each other. HD 148937 is continuously losing matter, which is being pulled away from its surface; occasional more violent outbursts produce the symmetrical shells shown in this photograph by David Malin, AAT.

# Contents

# Acknowledgments

There are many people who have helped me in the compilation of this book. First and foremost my thanks must go to Pieter Morpurgo, present Producer of the BBC 'Sky at Night' series, without whom the book would never have been written and to whom it is dedicated; grateful thanks too to Louise Tillard (Production Assistant), and Wendy Sturgess for her untiring research. During our travels round the world we were accompanied by a splendid BBC team: Jim Pearson and Kevin Baxendale (cameras), and Ron Keightley (sound), who were towers of strength.

We could have done nothing without the help of the astronomers and astronauts at the various observatories and NASA stations. My grateful thanks go to: Dr Arthur Hoag, Dr Charles Capen and Professor Clyde Tombaugh (Lowell Observatory and Las Cruces); Dr Edward Stone and Professor Harold Masursky (J.P.L.); Professor Geoffrey Burbidge, Dr Keith Pierce and Dr Jack Harvey (Kitt Peake); Professor Sir Fred Hoyle; Professor Sir Bernard Lovell and Professor F. Graham Smith (Jodrell Bank); Professor Malcolm Longair, Fred Watson and Dr R. Scobie (Edinburgh); Professor Antony Hewish and Dr Bruce Elsmore (Cambridge); Dr Terry Lee, Alan Pickup, Alfred Nield, Dr John Jefferies, Dr Dale Cruikshank and Professor René Racine (Mauna Kea); Professor Alec Boksenberg, Dr Paul Murdin and Joe Gietzen (La Palma); Dr John Ables and Dr Alan Wright (Parkes); Dr David Allen, Dr Keith Taylor, David Malin and Tom Cragg (Siding Spring); Dr Olin Wilson, Dr Douglas Duncan, Dr Robert Howard and Dr Graham Berriman (Mount Wilson); Sam Beddingfold (Cape Canaveral); Dr Herbert Beebe (Las Cruces); Commander Eugene Cernan (J.P.L.); Dr Ray Davies and Dr Keith Rowley (Homestake Mine Solar Observatory).

I have received every possible help from the publishers, with special thanks to David Roberts, Martin Bristow, Peter Coxhead and Caroline Zelkha. Barney D'Abbs and Iain Nicholson gave me great help with proof-reading. I am most grateful to Paul Doherty who provided the superb colour paintings specially for this book and to Hester Woodward for many valuable comments. To produce the book within a month of the 25th anniversary of the 'Sky at Night' programme in April 1982 was an achievement bordering on the miraculous!

Finally, last but by no means least, my thanks go to Hilary Rubinstein for his invaluable help.

PATRICK MOORE
Selsey, 1982 February 13

# Illustration Acknowledgments

The author and publishers are grateful to the individuals and organizations listed below for permission to reproduce copyright material:
The Anglo-Australian Telescope Board endpapers © 1981, frontispiece © 1981, 194 © 1977, 199 © 1977, 203 © 1980, 213, 227, 238; The Association of Universities for Research in Astronomy Inc 233, 237; Jack Bennett 138; BBC 14 (photo by Sam Andrew), 17 (photo by Pieter Morpurgo), 44 (photo by Kevin Baxendale); California Institute of Technology 181 © 1965; The Cerro Tololo Inter-American Observatory/The Association of Universities for Research in Astronomy Inc 204 © 1978, 234; Paul Doherty 134; Mary Evans Picture Library 112; Keystone Press Agency Ltd 18; The Kitt Peak National Observatory/The Association of Universities for Research in Astronomy Inc 178, 224, 230; The Lick Observatory 217; The Lowell Observatory, Flagstaff 70 *left*, 121, 126; Mauna Kea Observatory 125; Ludolf Meyer 130; Patrick Moore 10, 69, 70 *right*, 113, 118, 122, 123, 129, 144, 145, 152, 155, 157, 158, 160, 162, 166, 167, 168, 173, 175, 191, 195, 196, 197, 207, 208, 214, 221, 222, 226; Mount Wilson Observatory 139; NASA 8, 23, 24, 28, 29, 31, 33, 36, 37, 38, 40, 41, 42, 43, 47, 48, 49, 50, 53, 56–7, 58, 61, 63, 64, 66, 73, 75, 76, 78, 79, 80, 83, 84, 87, 88, 92, 94, 95, 96, 97, 98, 99, 102, 103, 105, 106, 108, 115, 119, 143, 151, 186, 188, 189, 244, 246, 247, 248; Novosti 20, 21, 27, 54; Palomar Observatory 136, 211, 212, 216, 229, 241; Royal Greenwich Observatory, Herstmonceux 243; Royal Observatory Edinburgh 13, 170 © 1979, 177 © 1980, 183 © 1974; Siding Spring Observatory/ Royal Observatory Edinburgh 169; Donald F. Trombino 149; USSR Academy of Sciences 45.

The author and publishers would particularly like to thank Paul Doherty for permission to reproduce the artwork on the following pages: 1, 34, 62, 82, 90–1, 100–1, 111, 115, 116–17, 118, 132–3, 140, 146, 164, 179, 190, 200, 215.

**Author's Note:** Here and there you will find references to 'The Sky at Night'. I like to believe that most people in Britain will know it, but others will not! It is a television programme, which I have presented monthly on BBC1 ever since April 1957 without a break; the first producer (without whom the programme would not have begun) was Paul Johnstone, who, sadly, died some years ago. Other producers who have been involved for protracted periods are Patricia Owtram, Patricia Wood and, today, Pieter Morpurgo. The 'Silver Jubilee' of the programme took place in April 1982, with a special programme entitled 'The Unfolding Universe'.

# Prologue

This is not a conventional book about astronomy. Neither have I set out to write a comprehensive text, or in fact a text of any kind. What I have tried to do is to tell the story of what has been going on during the past 25 years – the most exciting quarter-century in the whole history of science. My story is bound to have a personal slant, because throughout this period I have had the privilege of making monthly television commentaries, and doing my best to keep viewers abreast of the situation. When I began the 'Sky at Night' series in April 1957, I had no real idea of the startling developments that would take place before we reached our Silver Jubilee in April 1982.

Progress is only to be expected, of course, but in astronomy it has been almost explosively sudden. I have in my possession two copies of an astronomical book by Sir Robert Ball, one published in 1895 and the other in 1912, and when I compared them recently I found that there was remarkably little difference between them; astronomy at that time was more or less static. It is very different now, and a book published in, say, 1970 has become of historical interest only! The basic principles are the same, but the details alter almost from week to week.

Obviously the space programmes have had a great effect, but they do not represent the whole story, as was brought home to me in December 1981 when I visited Palomar Mountain, in California, during a round-the-world trip in connection with the 'Sky at Night' anniversary. Palomar, home of the world's most famous optical telescope, is some way from Los Angeles (though not far enough, as one astronomer wryly commented), and the drive up the steep road is pleasant; the domes at the top are impressive, and it was from here, from 1948 onwards, that astronomers peered further into space than had ever been possible before.

The Earth from space. This photograph was taken from the Apollo 11 spacecraft during its journey towards the Moon, carrying Neil Armstrong, Edwin Aldrin and Michael Collins. The spacecraft was 98,000 nautical miles from Earth when the picture was taken. Most of Africa can be seen, as well as parts of Europe and Asia. Many cloud structures are visible.

The Hale telescope, named in honour of the man who masterminded its construction, is a reflector; that is to say, it collects its light by means of a curved mirror, in this case 200 inches in diameter. Outwardly it does not look like the popular idea of a telescope, because instead of a solid tube it is a metal skeleton, mounted inside a gigantic horseshoe that can be driven round very accurately to follow the east-to-west drift of the stars across the sky. When the incoming light strikes the main mirror it is reflected up the skeleton tube, and is brought to focus, where an image of the object under study is formed. Originally, the observer stationed himself in a cage slung in the tube itself, and used cameras to take photo-

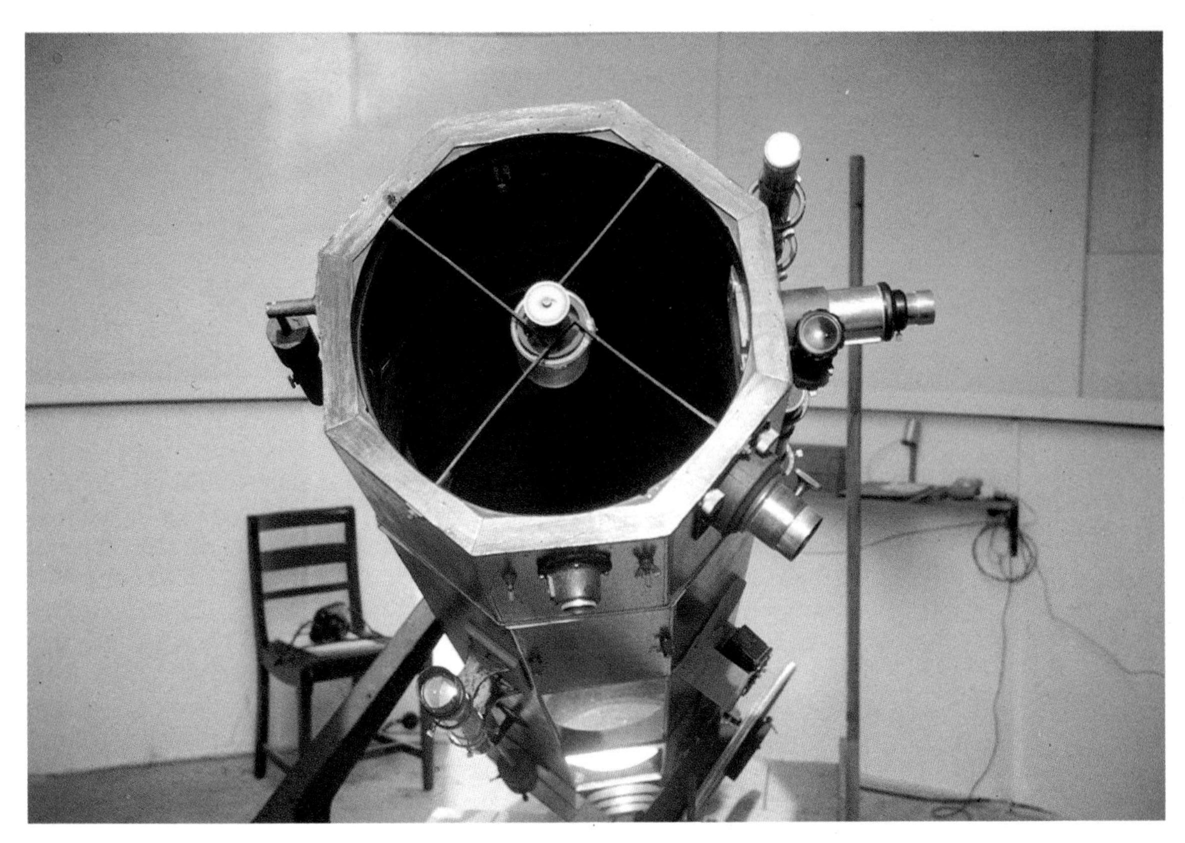

My own 15-inch reflecting telescope at Selsey in Sussex. The optical system is on the Newtonian pattern; the tube is wooden and octagonal, left partially open to prevent the formation of tube air-currents inside it. The telescope is housed in an observatory with a rotating roof, in which there is a slit so that the telescope can be pointed to any area of the sky. The mounting is equatorial, and of the 'fork' type; the driving mechanism is electric.

graphs of the faint, remote objects that the telescope was 'seeing'. Therefore, the observer was actually inside the telescope.

Reflectors of the size generally owned by amateurs are different, and instead of a cage there is a flat secondary mirror, placed at an angle so that it directs the light-rays into the side of the tube. With this arrangement, known as the Newtonian because it was devised by Sir Isaac Newton as long ago as 1668, the observer puts his eye – or his camera – to the side of the telescope. This is the case with the modest reflector in my garden at Selsey in Sussex, where the mirror is only 15 inches across instead of Palomar's 200, and I would have some difficulty in cramming myself inside the tube. (Another variant is the Cassegrain, in which the light is sent back from the secondary mirror through a hole in the main mirror, beyond which it is brought to a focus in the usual way.) But the procedure at Palomar now is very different from that of 1948, or for that matter 1957. Cameras are being replaced by electronic detectors, which are far more sensitive, and the observer no longer has to sit in a cage, checking the telescope patiently for hour after hour to make sure that it has not wandered away from its target. All the information is displayed on a television screen in the control room, just outside the main part of the dome, where the observer can sit in comfort, sipping his hot coffee. The telescope is guided by computers; there is nobody at all in the main dome.

I have chosen 1957 as my starting-point because it was then that the Russians launched the first spacecraft in history, the artificial satellite

Sputnik 1. It was small, and it stayed aloft for only three months, but in its way it was, I suppose, even more significant than the Palomar reflector, because it showed that man was capable of breaking free from the Earth. From a purely astronomical viewpoint, the most important aspect of space research is that observations can be carried out from above the top of the atmosphere, which makes a great deal of difference, so that without spacecraft our progress during the past 25 years would have been much slower than it actually was. But, before starting on my main theme I must pause for a few moments to set the scene, because there are some people who know nothing about astronomy, and I am anxious to make everything as clear as I possibly can. First, I must dispose of one misconception that still lingers on – the confusion between astronomy and astrology.

I have never been able to understand the confusion, because the two subjects are more different than the proverbial chalk and cheese. Astronomy is the science of the sky; it embraces everything from the nearby Moon out to the remote star systems far away in the universe. Astrology is not a science at all; it is the superstition of the sky, and its devotees claim that the apparent movements of the Sun, Moon and planets can affect human lives and destinies. When I was once asked to give my opinion about it, I made the reply attributed to the Duke of Wellington when asked if his name was Smith: 'If you will believe *that*, you will believe anything.'

So much for astrology. Now for a few bald facts.

The Earth is a planet, moving round the Sun at a mean distance of 93,000,000 miles. This may sound great, and by everyday standards it is; but it is not much to an astronomer, who has to reckon in millions of millions of miles as well as in immensely long periods of time. For instance, the Earth is over 4500 million years old, and the universe is much more ancient still. Nobody can really appreciate such figures, and the only course is simply to accept them. It is fair to add that astronomers cannot understand vast distances and time-spans any better than other people; the only difference is that they don't make the mistake of trying.

The Sun is an ordinary star – a globe of intensely hot gas, large enough to engulf more than a million Earths, and shining by its own energy, not because it is 'burning' but because of nuclear reactions going on inside it. It is, in fact, a huge nuclear bomb, though fortunately for us a controlled one. Round it move the Earth and the other planets, of which there are eight; Mercury and Venus closer to the Sun than we are, and Mars, Jupiter, Saturn, Uranus, Neptune and Pluto further away. The inner four, including the Earth, are comparatively small and solid, while the next four are large and differently constituted; gaseous outside, mainly liquid within. Pluto is a puzzle, and does not seem to fit into the general pattern, so that it may not be genuinely worthy of planetary status.

The Moon, magnificent though it looks, is a very junior member of the Sun's family or Solar System. It is the Earth's satellite, and keeps together with us as we move round the Sun. Like the planets, it has no light of its own, and depends entirely upon reflected solar rays. If for any reason the Sun were suddenly snuffed out, then the Moon and planets would

NGC 6559 and IC 1274-5. Rich region of the Milky Way, towards the direction of the centre of the Galaxy. The dark dust-lanes, bright reddish glowing gas patches and the soft blue light reflected from the edges of the clouds are well shown. These rims are shining by the light of nearby young, blue stars which have been formed in the surrounding dust-clouds, while the red-glowing gas is shining by its own light; it is made partly self luminous by the effects of the strong ultra-violet radiation from the same hot blue stars. The backdrop to this scene is formed by the huge masses of older, yellowish and much more remote stars which make up the central bulge of the Galaxy. The region lies in Sagittarius; the well-known Lagoon Nebula (M8) and Trifid Nebula (M20) are not far away from this complex of nebulæ. The photograph by David Malin, Anglo-Australian Observatory, was taken with the UK Schmidt telescope at Siding Spring.

vanish too, though the stars – suns in their own right – would shine on as before. (Luckily this is not likely to happen. The Sun will not change much for at least 5000 million years in the future; by stellar standards it is no more than middle-aged.) Other planets also have satellites, over 20 in the case of Saturn, and the Solar System is completed by bodies of lesser importance, such as the asteroids or minor planets, the comets, and assorted debris.

The Solar System is a small unit, no more important cosmically than a single grain of sand compared with the Sahara Desert. Even the nearest of the stars beyond the Sun is over 24 million million miles away, and to make calculations easier to handle we need a unit which is much longer than the mile or the kilometre. Accordingly, we make use of light, which flashes along at 186,000 miles per second. In a year it covers nearly six million million miles, and this distance, the light-year, is a convenient unit, so that the distance of the nearest star may be given as 4.2 light-years. The stars are suns, many of them much larger and hotter than ours, and they are so far away that they do not seem to move noticeably in relation to each other; the star patterns or constellations visible today are essentially the same as those that must have been seen in ancient times. It is only the members of the Solar System that wander about from one constellation to another.

The Galaxy, the star system that contains our Sun, is a flattened system about 100,000 light-years in diameter; I have rather unromantically likened its shape to that of two fried eggs clapped together back to back. When we look along the main plane of the Galaxy we see many stars in almost the same direction, producing the lovely Milky Way band so familiar to anyone who lives away from a city glaring with artificial lights. The Galaxy contains objects of many kinds, such as double stars, stars which vary in brightness, groups and clusters of stars, and clouds of gas and dust known as nebulæ, a few of which are visible with the naked eye.

Beyond the Galaxy there is a vast gap, and then we come to other galaxies so remote that their light takes millions, hundreds of millions, or even thousands of millions of light-years to reach us. They too contain their quota of stars, sometimes more than the 100,000 million suns of our own Galaxy; many of them are spiral in form, and some of them show evidence of violent internal activity. Still further away are the mysterious objects known as quasars, which were identified only in 1963 and about which our knowledge is woefully incomplete. The remotest known quasars may not be far from the boundary of the observable universe.

Note that when we look out into space, we are also looking backwards in time. Light takes 8.6 minutes to reach us from the Sun, so that we are seeing the Sun not as it is 'now' but as it was 8.6 minutes ago; the time-lag amounts to several hours for the outer planets, Jupiter and beyond, and to years for the stars, and even longer for the galaxies and quasars. Our view of the universe is always bound to be out of date, and even the stars in any particular constellation are not related to each other, because their distances from us are different, and once again we are dealing with nothing more than line-of-sight effects. Consider, for example, the two

Ursa Major, the Great Bear. From left to right the seven leading stars are: Alkaid, Mizar, Alioth, Megrez, Phad, Dubhe and Merak. Mizar is a naked-eye double; the small star close to it is named Alcor, while telescopically Mizar itself is seen to consist of two components. Megrez is decidedly fainter than the remaining six stars, and there have been suggestions that it has faded during the past 2000 years, though the evidence is very uncertain and it is not likely that any real change has occurred. Merak and Dubhe are known as the Pointers; they show the way to the Pole Star. Photograph by Sam Andrew

stars in the 'tail' of the Great Bear, the most famous of all the northern constellations. They are known as Mizar and Alkaid, and appear to lie side by side; but Mizar is 88 light-years from us, Alkaid as much as 210 – so that Alkaid is considerably further away from Mizar than we are.

How far does the universe extend? This is something we do not know. We may come to the limit of observation at between 15,000 and 20,000 million light-years, but there can be no actual boundary, so that either the universe is endless or else it curves back on itself, in which case the universe is finite but unbounded. This is not easy to explain in everyday language. I suppose the only possible analogy – and a very incomplete one, at that – is to imagine the attitude of a worm that has no idea of 'height', but somehow or other manages to crawl right round the Earth, finally coming back to its starting-point. To it, the world will then be finite and unbounded. With space we have to introduce further dimensions, but even mathematicians, who have a language of their own, cannot yet give a really clear answer.

Space research, at least for the moment, must be restricted to the Solar System. We can send men to the Moon and rockets to the planets, but to reach even the nearest star would require a travel time of many centuries, and I for one believe that if we are ever to undertake such journeys it will be by some fundamentally new method about which we know nothing as yet. Neither can we be sure that other stars are attended by inhabited planets, though most astronomers (not all) believe so. There is certainly no intelligent life in the Solar System except possibly on the Earth, so that we are very isolated in the universe.

From our exalted viewpoint of 1982, the astronomy of 1957 appears decidedly primitive. Since then we have developed new techniques and new theoretical methods; we have discovered objects of types not only unknown, but even unsuspected 25 years ago, and, of course, we have sent both manned and unmanned spacecraft into the Solar System. So let me now begin my main story by returning to Sputnik 1, and the real start of the Space Age.

# 1 Rockets from the Cape

I first heard about Sputnik 1 on 4 October 1957, when I tuned in to a BBC radio bulletin to be greeted with the news that the world's first artificial satellite was safely in orbit. I cannot honestly say that I was surprised, because it was an open secret that the Russians were preparing something spectacular, but I had not realized that they were so near to success. In this I was not alone. The American satellite programme, which had been announced almost two years earlier, had been beset with problems, and delay had followed delay. The reasons were partly technical, but also partly political, and the Russians had wasted no time.

Sputnik was football-sized, and carried little except a radio transmitter which sent out the never-to-be-forgotten 'bleep! bleep!' signals. It was moving around the world at a height ranging between 135 and 580 miles, completing one circuit in 96 minutes. As expected, it had been launched by a 'step-vehicle' involving three separate stages mounted one on top of the other, so that the uppermost stage, carrying the Sputnik, could be given what amounted to a running jump into space.

Reactions to the news were mixed. One earnest listener rang me at the BBC to say that he could confirm the flight of the Sputnik, because a vehicle from outer space had just landed in his back garden (I never found out exactly what it was). Others claimed to have seen brilliant lights in the sky, and even to have heard the strains of 'The Red Flag' wafting down from above. But the most popular question was: Could the Sputnik be seen – and if so, where?

In fact the Sputnik itself was never visible to the naked eye, because it was too small and faint, but the last stage of the launcher was larger, so that it outshone most of the stars when at its brightest. It could be recognized at once because of its apparently slow but steady movement against the starry background, and there was no real difficulty in locating it.

When I telephoned Moscow for the latest news, I was told that Sputnik was in a stable orbit, and there was no prospect of its coming down for several weeks. Rocket launching was the only possible method, because a rocket, unlike an aircraft, does not need to have atmosphere round it; instead it works on the principle of reaction, and, so to speak, 'pushes against itself'. The basic principle is the same as that of a firework rocket, where the hollow tube is filled with gunpowder. When lit, the gunpowder burns and sends out hot gas, which rushes through the exhaust in a concentrated stream and propels the whole vehicle in the opposite

Launch pad at the Kennedy Space Center, Cape Canaveral, Florida. It is from Canaveral that all the major United States space missions have been launched. This photograph was taken by Pieter Morpurgo in November 1981.

25 TON 10 TON

direction. Sputnik's launchers certainly did not use gunpowder, and instead of solid fuels there were two liquids, fed separately into a combustion chamber and reacting with each other to produce the necessary stream of gas.

Another question often put to me was: 'Why didn't the Sputnik fall down?' Again the answer was straightforward. Once a vehicle has been taken up above the atmosphere and put into a closed path round the Earth, travelling at sufficient speed (5 miles per second, or about 18,000 m.p.h.), it will remain orbiting, just as the Moon does. The only constraint is that the vehicle must remain well outside the denser part of the air, since if it has to push its way through the atmosphere it will be braked by friction, and will eventually spiral downwards, burning away in the lower air. The Earth's atmosphere extends upwards for many miles, but above a height of about 150 miles the density is so low that it has little effect upon a body moving through it. However, Sputnik's orbit took it regularly below this limit, and so it could not stay up indefinitely. I

Wernher von Braun, who was deeply involved in the United States rocket programme in its early days. He was one of the pioneers of the old German rocket society, which was taken over by the Nazis and transferred to Peenemünde in the Baltic; here the team led by von Braun produced the V2 rocket weapons which may be regarded as the direct ancestors of the space vehicles of today. Subsequently von Braun went to America, and remained active in research until his death in 1977. It is fair to say that he was always far more interested in space flight than in weapons of war!

estimated that it would last between two and three months, and in this I was right; Sputnik came to the end of its career during the first week of 1958.

Meanwhile the Americans had not been idle. They had set up a rocket ground at White Sands in New Mexico, in the foothills of the Organ Mountains, and had carried out a long series of tests with the intention of launching an artificial satellite before the end of 1958. One of their leading technicians was Wernher von Braun, who had masterminded the V2 rocket campaign against England during the closing stages of World War II, and had subsequently gone to the United States to design rockets of a more peaceful nature. I first met von Braun in 1952, and later came to know him well; I am convinced that if he had been given a free hand, the first artificial satellite would have come from the U.S.A. instead of from the U.S.S.R. Unfortunately there was a good deal of inter-service rivalry between the army and the air force, and it was not until February 1958 that von Braun's team successfully dispatched America's pioneer satellite, Explorer 1. It was a mere six inches in diameter, but it proved to be far more valuable scientifically than Sputnik, because it led to the first of the great discoveries made soon after the Space Age opened.

Explorer carried more than a radio transmitter. It was also equipped with an instrument known as a Geiger counter, used to detect electrically charged particles and high-energy radiation. The main programme was to study cosmic rays, which are not really rays at all, but nuclei of atoms that come from space and bombard the Earth constantly. The relatively heavy nuclei are broken up when they plunge into the upper air, and only their fragments reach the ground. In 1958 the origin of cosmic radiation was a mystery, and in fact we still do not know the whole story. Some cosmic rays come from the Sun, but most of them originate far beyond the Solar System, and they may well pervade the whole of the Galaxy.

Explorer's Geiger counter was expected to show an increase in the numbers of cosmic-ray particles at high altitudes, and at first this is what happened, but at a height of 600 miles the cosmic radiation counts suddenly stopped. Explorer was travelling in a more eccentric orbit than Sputnik 1, ranging between 224 miles and as much as 1585 miles above the ground, so that it went through the 600-mile limit at every circuit. The investigators, led by James Van Allen, were frankly puzzled. It seemed absurd to suggest that there could be no cosmic-ray particles above 600 miles, but none could be recorded.

It took much hard work, and several more cosmic-ray satellites, to find the answer. The Geiger counters had 'gone dead' at 600 miles not because there was too little cosmic radiation, but because there was too much. The counters were saturated, and were unable to cope. Today we know that there are two main Van Allen belts, arranged in doughnut form round the Earth; they are due chiefly to electrically charged particles sent out by the Sun (the so-called solar wind) and then trapped in the Earth's powerful magnetic field.

Sputnik and Explorer showed that satellites had come to stay, and during the next four years there were many launchings, both Russian

Launching of the first manned space flight: Vostok 1, carrying Yuri Gagarin, on 12 April 1961. Reports that the Russians had made earlier, unsuccessful attempts are certainly untrue.

and American. Now, of course, Earth-orbiting vehicles are so important that it is difficult to imagine the world without them; to give only one example, it would be strange to lack direct television links between Europe and the United States.

Another highlight was that of 12 April 1961, when Major Yuri Gagarin, of the Russian Air Force, went up in the spacecraft Vostok 1, and made a complete circuit of the Earth, well above the main part of the atmosphere, before landing safely in the prearranged position. He was aloft for only 1 hour 48 minutes, but his flight was profoundly significant, because up to that time there had still been many pessimists who believed that manned space travel would never be possible.

Naturally, I was anxious to meet Yuri Gagarin, and some months later I was able to do so, when I attended a scientific conference in Moscow. Gagarin was there, and quite ready to talk. My knowledge of Russian is nil, but luckily his English was reasonably fluent.

My first question was perhaps rather too obvious: 'Were you badly frightened when you were lying in that cramped capsule, waiting to be blasted off into space?' He was quite honest: 'I might have been, but there was no time to think of anything except the mission itself' – a very logical answer.

I then asked him about the various hazards which had always been regarded as serious. First there is weightlessness, or zero gravity. Once an astronaut (or cosmonaut, to use the Russian term) is moving round the

Gagarin, during his space flight. This was the first time that a man had been subjected to zero gravity over a prolonged period.

Earth in free fall, he has no apparent sensation of weight, not because he has escaped from the Earth's gravitational pull but because he and his vehicle are moving in the same direction at the same rate, and there is no pressure from the one to the other. This is something which cannot be satisfactorily reproduced on Earth, except for very brief periods in a diving aircraft, so that the effects had been impossible to predict. It had been suggested that an orbiting astronaut would become disoriented at best, violently sick at worst. Gagarin assured me that such fears had been groundless. 'In fact,' he said thoughtfully, 'I found weightlessness a very pleasant experience.'

Later astronauts have confirmed this, and we now know that zero gravity is not harmful over limited periods. However, we must not be too complacent, and it is by no means certain that the human body will be able to cope with really long space flights of the type needed to reach Mars or Venus, but the situation is brighter than it had seemed before Gagarin's mission. Another problem which has proved to be less serious than expected is that of meteoroids – solid particles in orbit round the Sun. The small ones are too fragile to do much damage to a spacecraft, and evidently the larger bodies are scarce, though I suppose we may have to reconcile ourselves to occasional disasters when space travel becomes really commonplace.

I liked Yuri Gagarin. He was an excellent publicity man as well as an excellent cosmonaut (no doubt one of the reasons why he had been chosen), and he fully expected to make further flights. When I asked him whether he hoped to go to the Moon, he replied: 'If I am selected, there will be no difficulty.' Unfortunately it was not to be; Gagarin was killed in an ordinary aircraft crash some years later. History will never forget him.

Before the end of 1961 another Russian cosmonaut, Gherman Titov, had made a flight of longer duration than Gagarin's, and it was then, I think, that the idea of a 'space race' began to be widely discussed. There seemed to be a general impression that the U.S.A. and the U.S.S.R. were taking part in a scramble to reach the Moon. Moreover, it is true that rockets used to launch spaceships can also be used for military purposes, and it was clearly desirable to have orbiting Americans as well as orbiting Russians. In America the centre of interest shifted from White Sands to the new testing range at Cape Canaveral, in Florida.

The Cape is particularly suitable, because the initial stages of the rockets launched from there can fall harmlessly in the sea. When I first went to the Cape, it was a scene of tremendous activity, and there was great enthusiasm – not shared, I imagine, by the local armadillos and occasional alligators which had lived there long before the rocket technicians had arrived. During the 1960s the site's name was altered to Cape Kennedy, but the change was not popular, and after a few years it was dropped.

There is much to see at the Cape. I have always been particularly impressed with the VAB or Vehicle Assembly Building, in which various important spacecraft have been assembled (the latest being, of course, the Shuttles). The VAB is one of the largest single buildings in the world, with a height of over 500 feet and a floor area of eight acres. I have been told that clouds can form on the roof, and that spots of rain have been known to fall even when the weather outside has been beautifully clear and sunny, though whether or not this is true I have never been able to decide! Another feature of the Cape is the huge crawler transporter, built to take vehicles from the VAB to the actual launching sites. The crawler is over 130 feet long, and since its maximum speed is only two miles per hour it is probably the world's slowest vehicle. It may also be the heaviest; a specially strengthened road had to be built for it.

All the important American space missions, both manned and unmanned, have been launched from Canaveral. Inevitably there have been failures. Early rockets were not reliable, and there was one which caused an international incident when it crash-landed in Cuba and killed a cow – which was promptly given a State funeral as a victim of Imperialist aggression. But by 1961 the worst of the problems had been overcome, and it is significant that no American astronaut has ever been killed during a mission. Let us hope that this record will be maintained.

I had not heard about Gagarin's flight until after it was over, since the Russians had made no announcement, but I and everyone else knew about the first American mission, that of Commander (now Admiral) Alan Shepard. It took place on 5 May 1961, less than a month after Gagarin's, and was less ambitious inasmuch as Shepard did not enter orbit. He made an up-and-down 'hop', reaching a peak altitude of 116 miles; the trip lasted no more than a quarter of an hour, and it was broad-

Commander (now Admiral) Alan Shepard was America's first man in space. Subsequently he was taken off the active list on medical grounds, but remained as a leading member of the organization, and was involved in the training of future astronauts. Eventually he was restored to the active list, and in 1970 commanded the Apollo 14 mission to the Moon. He is shown here together with the Apollo 14 emblem.

cast to the world, so that from my Sussex study I could actually hear Shepard talking to ground control. It would have been difficult then to believe that less than ten years later, Shepard would be able to step from a spacecraft on to the surface of the Moon.

As the decade passed by, so the space programmes, both American and Russian, became more reliable and also more scientifically valuable. The one-man (or, in a single case, one-woman) capsules were succeeded by larger vehicles capable of carrying two or three passengers; it became common for two vehicles to meet in space and dock together, and astronauts were able to go outside their vehicles, achieving what is usually though misleadingly called spacewalking. Sadly, there were tragedies too; three United States astronauts were killed during a rehearsal on the ground, when their cabin caught fire, and there were four Soviet casualties during actual missions. These were grim reminders that space is a hostile environment, and that travel beyond the Earth is always a risky business. But despite these disasters, the programmes continued – and the next goal was our sister world, the Moon.

# 2 Reaching for the Moon

There was never any doubt that the Moon would be our first target, simply because it is so close to us. Its distance is on average a mere 239,000 miles, as against well over 20,000,000 miles for even the nearest planet, Venus. Moreover the Moon is our companion, and orbits the Sun together with us, so that a rocket flight there and back can be completed in less than a fortnight.

Yet the Moon has its disadvantages, the worst of which is the total lack of atmosphere. Even if the Moon had any 'air' in its youth (which is doubtful) it has none today; the gravitational pull is so weak that no atmosphere can be held down. Without air there can be no water, and no life.

The surface features are easy to see even with the naked eye, and any small telescope will show a mass of detail. There are broad dark plains, miscalled seas, which have been given romantic names such as the Mare Imbrium (Sea of Showers), Oceanus Procellarum (Ocean of Storms) and Sinus Iridum (Bay of Rainbows); there are mountains, hills and valleys, and there are thousands of walled enclosures known as craters, ranging from tiny pits up to vast structures well over 100 miles in diameter, with sunken floors and, in many cases, high central peaks. But at the start of the Space Age our information was by no means complete, and several vital questions had to be answered before any attempt could be made to send an astronaut to the lunar surface.

For instance, was the ground firm enough to bear the weight of a spaceship? According to one theory, the 'seas' were likely to be filled with soft dust, so that any vehicle unwise enough to land there would be promptly and permanently swallowed up. Telescopic observation could not decide one way or the other; the only course was to send unmanned probes to find out.

This is precisely what the Russians set out to do in 1959, with their first three 'Luniks'. Obviously, these experiments were preliminary only, but several important discoveries were made, and the results were highly encouraging.

Lunik 1 passed within 4660 miles of the Moon in January, and sent back data of real interest; in particular, the Moon appeared to have no measurable magnetic field. Lunik 2, dispatched in the following September, crash-landed in the grey plain of the Mare Imbrium, so making the first direct contact between Earth and Moon even though it did little else. But it was Lunik 3 which really stole the limelight, because it sent

The Moon from Orbiter 4. The photograph was taken on 11 May 1967, when Orbiter was 1850 miles above the lunar surface. North is at the top, and sunlight is coming in from the right. Most of the area shown here is on the far side of the Moon, and therefore never visible from Earth, though the long valley associated with the walled plain Schrödinger, near the bottom of the picture, is only just out of view beyond the Moon's south pole. The dark-floored feature Tsiolkovskii is visible near the right-hand edge.

OPPOSITE: The first picture from Lunik 3, October 1959. The solid line crossing the Moon is the lunar equator, while the dotted line indicates the boundary between the visible and averted parts of the Moon as seen from Earth. Various features are identifiable. Roman numerals indicate objects on the Earth-turned region: **I** Mare Humboldtianum (Humboldt's Sea), **II** Mare Crisium (Sea of Crises), **III** Mare Marginis (Marginal Sea), **IV** Mare Undarum (Sea of Waves), **V** Mare Smythii (Smyth's Sea), **VI** Mare Fœcunditatis (Sea of Fertility) and **VII** Mare Australe (Southern Sea). On the far side, and shown here for the first time, are: **1** Mare Moscoviense (Moscow Sea); **2** Bay of the Astronauts; **3** continuation of the Mare Australe; **4** Tsiolkovskii; **5** Lomonosov crater; **6** Joliot-Curie crater; **7** Soviet Mountains; **8** Sea of Dreams: By modern standards the picture is poorly defined, but it must be remembered that the sunlight was coming almost straight down on to the area being photographed. The main error in interpretation concerns the Soviet Mountains, which do not exist at all; the name has been deleted!

back pictures of a region previously absolutely unknown – the Moon's far side, which is always turned away from us.

The Moon takes 27.3 days to complete one journey round the Earth. It spins on its axis in exactly the same period, so that the same hemisphere faces Earth all the time, while the opposite hemisphere is never visible from our planet. (The best way to show what is meant is to walk slowly round a chair, turning so as to keep your eyes chairward; anyone sitting on the chair will never see the back of your neck.) Theorists could explain this behaviour easily enough; it is due to tidal friction over the ages, and most of the satellites of other planets have similarly captured or synchronous rotations, but it was irritating to be so ignorant about a large portion of the nearest object in the sky. Actually, we can examine rather more than half the total surface because of what are termed librations; the Moon's orbital velocity varies somewhat, and the position in orbit and the axial spin become periodically out of step, so that we can see first a little beyond one mean limb and then beyond the other. In fact, the Moon seems to wobble slightly to and fro, though the regions close to the edge of the disk are always very foreshortened, and are difficult to map. But before 1959 we had no knowledge at all of the permanently averted 41 per cent.

Lunik 3 altered all this. It was launched on 4 October 1959, exactly two years after Sputnik 1, and it went on a round trip. By 7 October it was 35,000 miles on the far side of the Moon; its cameras were switched on by remote control, and after the photography had been completed the films were automatically developed and fixed. Then, while the Lunik drew back towards Earth, the pictures were transmitted by television, and were safely received by the ground-based Russian operators.

I was particularly interested in this, because I had spent many years trying to map the foreshortened limb regions of the Moon, and I had my own ideas about what the far side would look like. My first view of the pictures was during a 'Sky at Night' programme; I was 'on the air' when the producer told me that the Lunik pictures were about to appear on the screen, and I remember wondering whether I would be able to make head or tail of them. I hoped for the best, and waited in a state of considerable excitement.

Then the first picture came up. For a few moments I stared at it, and by sheer luck, I recognized a feature which I knew: the well-defined dark plain of the Mare Crisium (Sea of Crises), seen under an unfamiliar angle of illumination but still identifiable. The rest of the picture was lacking in well-marked detail, but there were two clearly visible dark patches and a bright streak which was at first (wrongly) believed to be a mountain range.

The one obvious fact was that the far side of the Moon differed from the near side in one important respect – there were no large 'seas'. Later, when I was able to take a more leisurely look at the pictures, I could make out a few craters, but little else. By modern standards the Lunik results were very poor, though in 1959 they represented a tremendous achievement.

I expected the Russians to follow up their success quickly, but they did not, and in fact their next probe, Luna 4, of April 1963, was a failure,

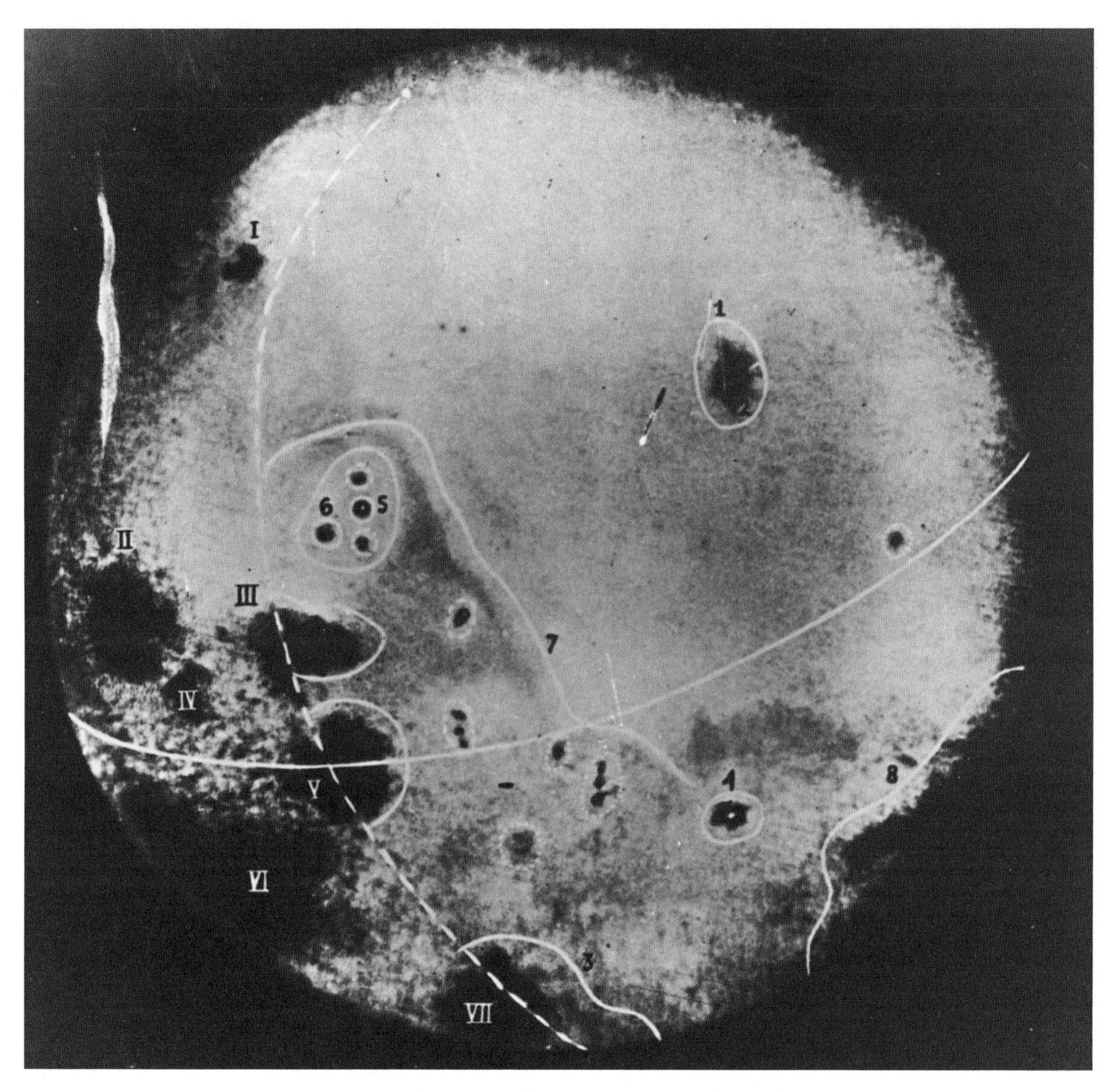


missing the Moon by over 5000 miles. The next major advance came from America, with the probes of the Ranger series. This time the plan was to crash-land on the Moon's surface, all the pictures being taken and sent back during the last few minutes of the flight. For various reasons, the first six Rangers failed, but on 31 July 1964 the seventh impacted on the Mare Nubium (Sea of Clouds) after having transmitted more than 4000 pictures.

For the first time we had a really detailed close-range view of the Moon's surface. As expected, there was very little level ground anywhere, and even the apparently smooth areas were pitted with craters, hummocks, ridges and minor hills. There was plenty to see, but everything emphasized the stark, hostile nature of the lunar surface.

Rangers 8 and 9 followed in early 1965. The latter was of special interest, because it came down on the floor of the large, walled plain,

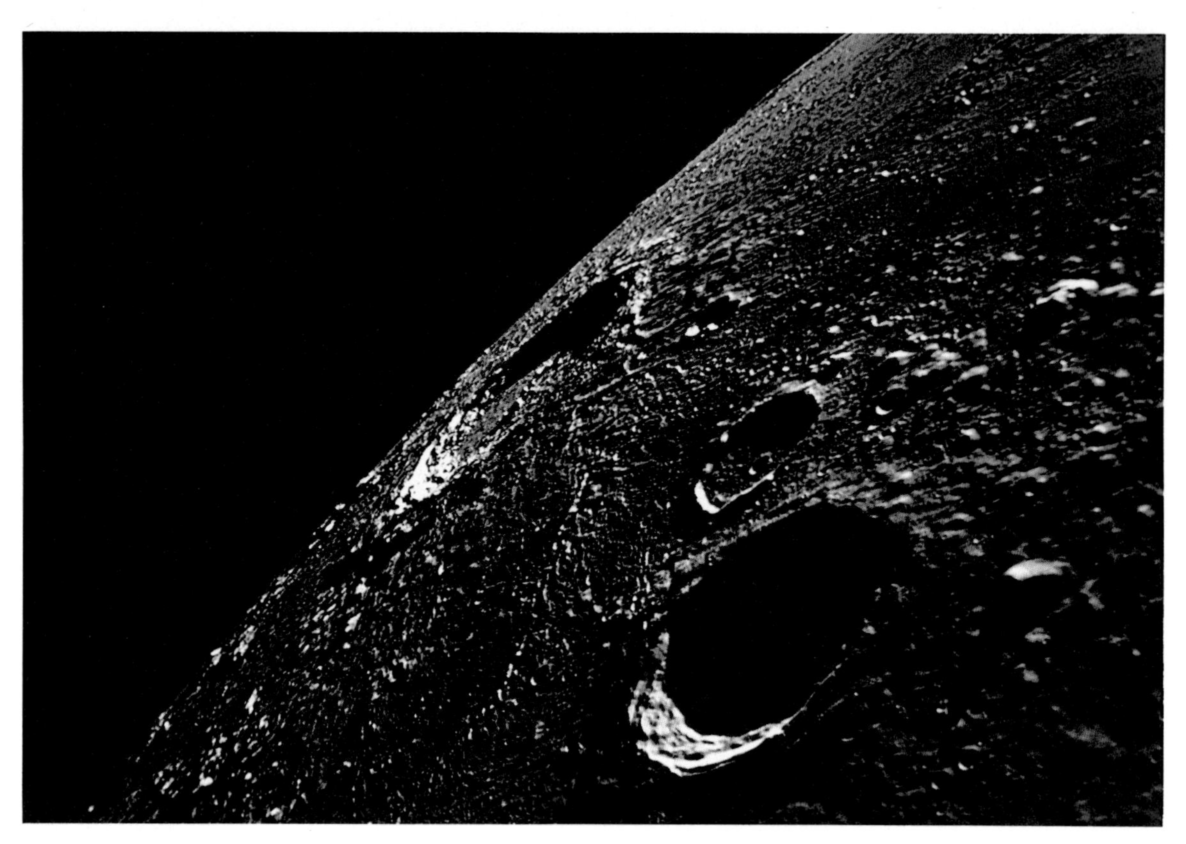

The Moon from Apollo 12. At the back appears the great ray crater Copernicus, one of the two major ray centres on the Moon; from Earth it is very prominent, with high, terraced walls and a central mountain complex, but here it is shown as very foreshortened. The crater Reinhold, well seen in this photograph, is 30 miles in diameter, with walls which rise in places to 9000 feet above the sunken floor. Adjoining Reinhold, more or less between it and Copernicus, is the 15-mile crater Reinhold B.

Alphonsus, where local activity had been suspected; obscurations and reddish glows had been seen there, due presumably to gas seeping out from below the surface. Ranger 9's pictures showed no traces of activity, but they did provide us with an excellent view of the crater floor with its mountains, its ridges, and its complicated system of cracks or rills.

So far, so good – but what about the all-important question of the surface strength? If the soft-dust theory were right, then manned journeys would probably be impossible. And so it was a great relief when the next successful Russian probe, Luna 9, showed otherwise. This time the landing was controlled, so that as the vehicle neared the Moon it was braked by rocket power, and touched down gently enough to avoid serious damage. Within a few minutes of its arrival pictures were being received, both in the Soviet Union and at Jodrell Bank in Cheshire, where the great new radio telescope had just been brought into use. Luna 9 did not sink into dust. It stood on perfectly firm ground, and the general scene reminded me strongly of an Icelandic lava plain, with rocks and boulders everywhere. One rock cast a long, pointed shadow which looked enormous, even though the rock itself can hardly have been more than six inches across.

Space planners everywhere felt elated: if we set out to reach the Moon at least there would be no fear of being swallowed up. Luna 9 had done its work well. Before long its power failed, but we know exactly where it is, and no doubt it will eventually be collected and taken to a

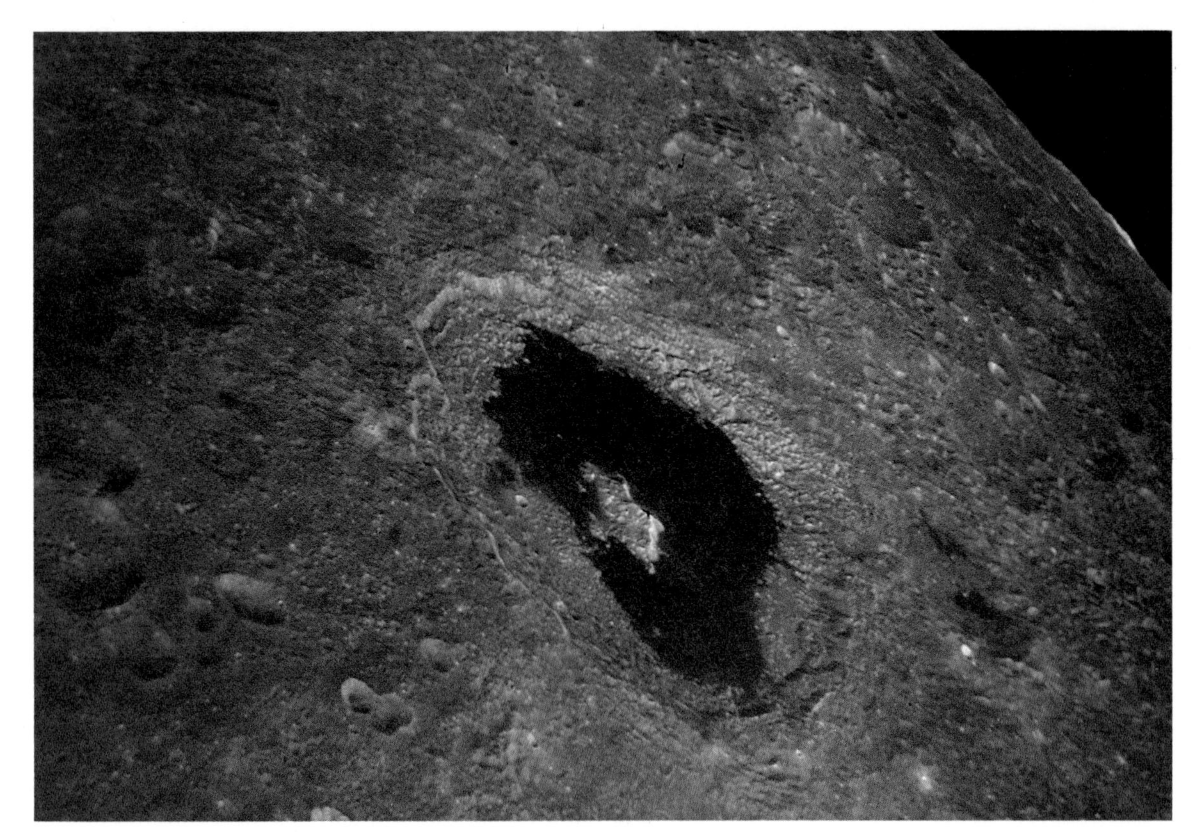

Tsiolkovskii. This great formation is 318 miles in diameter, much larger than any walled plain on the Earth-turned side of the Moon. It intrudes into another formation, Fermi, which is of comparable size but does not have a dark floor. The apparent blackness of the interior of Tsiolkovskii shown here gives a false impression of shadow.

museum. On the Moon, remember there is no weather, and there is nothing to damage Luna 9.

As so often happened in those pioneering days, a Russian success was followed soon afterwards by an American one on the same lines. This time it was the first of the Surveyor probes, which came down about 500 miles from Luna 9 and sent back more than 11,000 pictures; it even survived the intense cold of a lunar night, and began transmitting again after sunrise. Overall the pictures were better than those from Luna 9, and there were more of them, but the general aspect of the landscape was much the same.

It was something of an anti-climax when the next Surveyor, launched from Canaveral in September 1966, went out of control and crashed on the Moon without sending back any data, but, as one of the team said, it would be conceited to expect that 'the jackpot could be won every time'. More successful probes followed, one of which, Surveyor 7, came down on the wall of one of the Moon's most spectacular craters, Tycho in the southern uplands.

Tycho is of special note because it is the focal point of a system of bright rays, which spread out in every direction for hundreds of miles. They are not visible under low angles of illumination, but near the full moon they dominate the scene, and even large craters become hard to identify. There are other important ray systems, notably that of the 56-mile crater Copernicus in the Mare Nubium, but none can rival Tycho's.

Obviously, we want to find out how the craters on the Moon were formed, and at present there are two opposing schools of thought. The craters may be volcanic, in which case they will be of the same type as terrestrial calderas. Or they may have been produced by an intense bombardment by meteorites in the early part of the Moon's history. No doubt both processes have played their part, and at least we have a good idea of the time scale; the principal basins were flooded with lava around 4000 million years ago, producing the seas or maria of today. There are signs of lava flows everywhere, and craters on the edges of the seas have had their walls breached or even levelled.

The craters are not spread at random all over the Moon; there are certain definite laws. When one crater breaks into another, it is almost always the larger formation that is distorted, and the walled formations of all sizes tend to make up groups or chains. The Moon has had a violent career, but it is, to all intents and purposes changeless now, and no major craters can have been formed for at least the past 1000 million years.

Tycho must be one of the youngest of all, because its rays pass over all the surrounding features. Surveyor 7 showed its outer slopes, covered with massive rocks and boulders, making it a most unsuitable site for a manned landing.

Luna 9 and the Surveyors settled the doubts about the firmness of the Moon's crust, but it was still only too likely that there would be unsafe areas here and there, and before sending any manned expedition it was absolutely essential to have detailed maps. Earth-based observations were not accurate enough, and so the Orbiter probes were dispatched to complete the task that had been begun so long before.

All the five Orbiters were completely successful. Between August 1966 and the end of 1967 they were put into closed paths round the Moon, and sent back so many thousands of photographs that even today not all of them have been thoroughly studied. The amounts of detail shown were truly remarkable. The differences between the near side and the far side were confirmed; not only does the far side lack any large 'seas', but the arrangement of its craters is not exactly the same, and it has some peculiar features not to be found elsewhere. One of the most notable of these has been named Tsiolkovskii, in honour of one of the early Russian space flight theorists. This striking feature was even shown on the first Lunik 3 picture; it has high walls, a dark floor that is unquestionably a dry lava lake, and a lofty central peak, so that it seems to be intermediate in type between a crater and a sea.

As the Orbiters travelled round and round the Moon, it was discovered that their movements were not quite regular. Sometimes they speeded-up slightly, and sometimes they slowed down – not by much, but by a measurable amount. This behaviour was tracked down to blobs of dense material below the Moon's crust, now termed mascons, a convenient abbreviation for *mas*s *con*centrations; before an Orbiter passed above a mascon it was accelerated, and subsequently held back. It was suggested that mascons might be buried meteorites, but this does not seem to be at all likely. More probably the mascons are disk-shaped areas of volcanic rock not far below the surface.

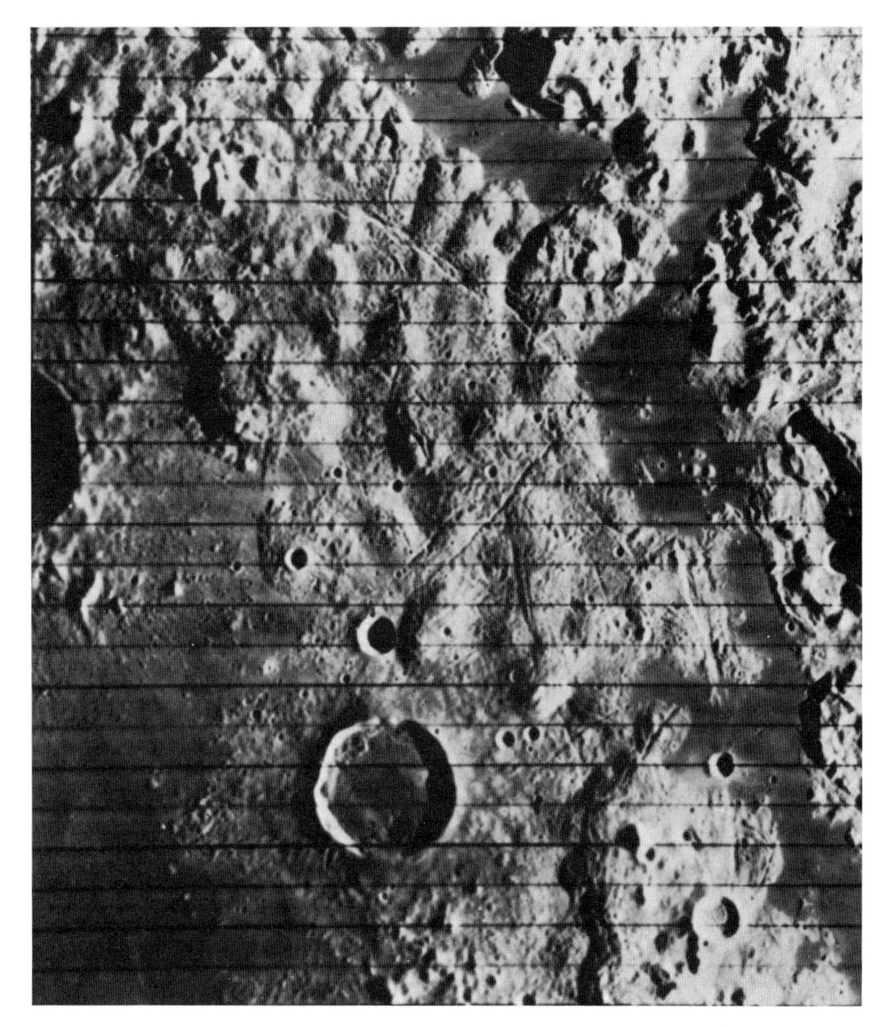

Part of the Mare Orientale (Eastern Sea), photographed from Orbiter 4 in 1967 together with the mountainous border, which is just visible from Earth and makes up what are called the Rook Mountains. The Mare Orientale itself is so foreshortened as seen from Earth that it is not easy to identify, and it was not until the space probes sent back their results that it was found to be a vast, multi-ringed structure; it is, incidentally, associated with a mascon. I suggested calling it the Eastern Sea because it lay close to what was then taken to be the eastern limb of the Moon. Subsequently a decision of the International Astronomical Union reversed east and west, so that the *Eastern* Sea is now at the Moon's *western* limb. I opposed the change, but was heavily outvoted!

One of the mascons lies below the feature called the Mare Orientale (Eastern Sea), a vast ringed basin that extends to the Moon's far side. From Earth it can hardly be recognized, and it had not been mapped before I first drew attention to it in 1946; it is so close to the edge of the disk, that even when best placed, it is very foreshortened. When I discovered it (and, incidentally, suggested its name) I had no idea that it would prove to be a major feature, but the Orbiter photographs showed it in all its glory.

Many other lunar probes were sent up between 1959 and 1969, some of which even carried out preliminary analyses of the surface rocks and showed them to be of volcanic type. Meanwhile the Apollo programme was working up to its climax, and the first 'Moon-men' were being trained.

There have been many suggestions that the Russians, too, intended to reach the Moon before 1970, but I am dubious; it seems much more likely that the two space programmes were running along different lines. I may be wrong, but today it really matters very little. What is important is that in July 1969 the first man stepped out on to the lunar surface.

# 3 Apollo

I first saw an Apollo spacecraft in the summer of 1968, when I went to Cape Canaveral after the end of a scientific conference. Everything seemed to be going according to schedule; Apollo 8 was almost ready, and on the following 21 December it was due to take Astronauts Frank Borman, James Lovell and William Anders on a flight round the Moon. The launcher was to be a Saturn rocket, the most powerful ever built. It towered to a height of 360 feet, equal to that of St Paul's Cathedral – and yet only the top cone, a mere 22 feet long, would survive to bring the astronauts home when their mission had been completed.

Two different plans had been considered for the pattern of the journey. One involved putting the whole vehicle into a closed orbit round Earth, and launching the lunar probe from there. Wernher von Braun told me that this was his first idea, though later he changed his mind to the plan that was adopted. This was to use the immense power of the Saturn rocket to send the actual probe, made up of a Command Module and a Service Module, towards the Moon – together with the spidery, flimsy-looking Lunar Module. For the landing itself, two of the astronauts would go down in the Lunar Module, and then use the single ascent engine to re-join the third member of the crew still in the main part of the vehicle. The Lunar Module would then be jettisoned, and the return journey could begin.

It seemed a clumsy procedure, but there was no practical alternative, and the decision turned out to be a wise one. Preliminary tests had been carried out with Apollo 7, but only at an altitude of a few hundred miles; Apollo 8 was to be the first manned vehicle to approach the Moon.

There were no hitches. On 21 December the astronauts blasted off, and by Christmas Eve the Apollo was near the Moon, ready for what was termed 'injection into lunar orbit'. At Mission Control, in Houston, this was one of the most anxious moments of the entire flight, because the manœuvre had to take place with Apollo 8 on the far side of the Moon and therefore completely cut off. If the firing had been faulty, and the spacecraft had crashed to destruction, nobody on Earth would ever have known just what had gone wrong.

Going down on to the Moon; the Lunar Module, carrying the two moon-walkers, during its descent. The Earth appears as a tiny crescent.

Fortunately all went well. Apollo came round at the expected moment; I was making a broadcast at the time, and I still recall my sense of relief when I first heard Colonel Borman's voice reaching me across a quarter of a million miles of space. The vehicle entered an elliptical path taking it from 69 to 195 miles above the lunar surface, but two orbits later the

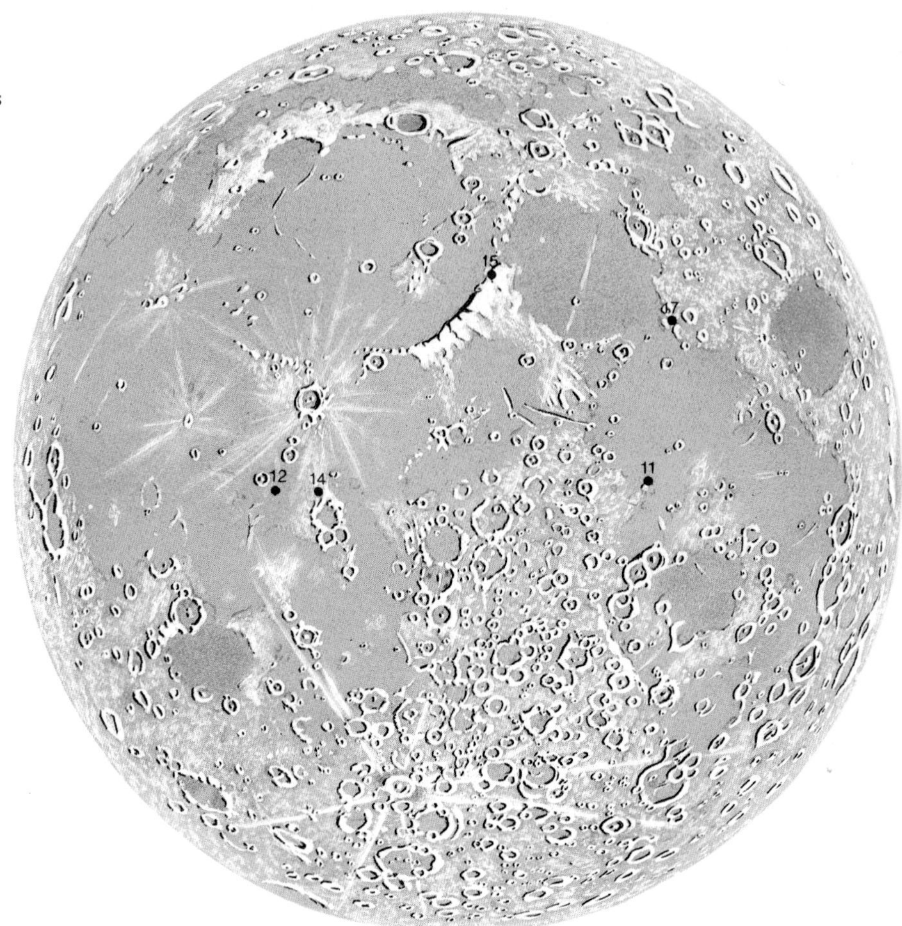

Map of the Earth-turned hemisphere of the Moon, with the positions of the six Apollo landing sites. North is at the top. Drawing by Paul Doherty

motors were fired again, putting Apollo into a nearly circular path 70 miles from the Moon.

For almost a day, Apollo 8 remained in orbit. Apart from the dead periods when the spacecraft was behind the Moon, communications were good all the time. 'The colour of the Moon looks a very whitish grey,' commented Anders. 'Some of the craters look like pickaxes striking concrete, creating a lot of fine haze dust . . . The sky is pitch black. The contrast between the sky and the Moon is a dark line.' The descriptions were fascinating to hear; after all, the lunar craters were being seen from close range for the first time.

On Christmas morning, with Apollo 8 once more behind the Moon and completely out of touch, the motors were fired again, breaking the spacecraft free from the Moon's grip and putting it into a homeward path. The journey ended on 27 December, with a perfect splashdown 1000 miles southwest of Hawaii.

The first expedition had been carried out entirely according to plan. Apollo 9 came next, testing the Lunar Module in an orbit round Earth, and then everything was ready for the dress rehearsal – Apollo 10, carrying Thomas Stafford, Eugene Cernan and John Young. They made a perfect crew. All had been in space before (Stafford twice, and the others once each); they knew what to expect, and they not only achieved everything they had hoped to but also did their utmost to share their experiences with the world. They revealed a tremendous sense of humour, and they took an obvious pleasure in putting on what might be termed 'displays' for the benefit of the millions of people watching them on television.

During one of my television broadcasts, I remember saying that Stafford, Cernan and Young had set a standard that future crews would find hard to match. And yet, ironically, their flight is barely remembered today except by those who watched it – because they 'only' went to within ten miles of the Moon, surveying the proposed landing-site for the next mission and testing the Lunar Module to its utmost. They finally splashed down in the Pacific on 26 May 1969.

And so – on to Apollo 11, which will never be forgotten as long as humanity survives. Two of the astronauts were scheduled to land on the Moon, Neil Armstrong first and then Edwin Aldrin, leaving the third member of the team, Michael Collins, circling in the Command Module. What they had to do was far more dangerous than anything attempted before. They had to land the Lunar Module, *Eagle*, on the Sea of Tranquillity, go outside for a period of over two hours, and then blast back to rendezvous with Collins. If either of the *Eagle*'s motors faltered, the result would be disaster.

Apollo 11 was launched on 16 July, and swung out towards the Moon in the usual way. On arrival it was put into a circular path 70 miles above the lunar surface, and then, at last, the final phase began. Armstrong and Aldrin went into the Lunar Module, separated it from the main spacecraft, and dropped towards their goal. As they neared the Moon their calm voices came through to Mission Control, to be relayed across the world; the braking rockets were fired, and *Eagle* hovered a mere 150 feet above the surface. Neil Armstrong took over manual control. In such a situation, no computer can be as good as the human brain.

*Eagle* had only enough fuel for two minutes' hovering, but as Armstrong tilted the Module he could see that directly below there was a crater 'about the size of a football pitch', with unpleasantly large rocks strewn all across it. He made a quick decision, and manœuvred *Eagle* down-range towards smoother ground. Gently he brought the spaceship down until the wires projecting from the footpads touched the surface. A second later he cut the engines, and *Eagle* came to rest.

It was a tremendous moment. I heard Neil Armstrong's words: 'The *Eagle* has landed'. What I said to the television audience I do not remember; I only hope that it was appropriate to the occasion. The time was a few hours before midnight on 20 July.

According to the original flight plan, the astronauts were due for a 'sleep' period before going out of the Module. Armstrong and Aldrin had other ideas, and since they were so obviously in good shape there was no reason to delay. At 1.53 G.M.T. on 21 July the cabin was depressurized. Half an hour later all was ready; the hatch was opened, and Armstrong crawled out on to the ladder. By 2.55 he had reached the footpad, and paused. His words came through clearly:

'I'm at the foot of the ladder. The LM footpads are only depressed in

Apollo 11; the Command and Service Modules, photographed from the Lunar Module while in orbit round the Moon. The lunar surface below is the region of the Mare Fœcunditatis (Sea of Fertility). About half of the crater Taruntius G is visible at the lower left of the picture, and part of Taruntius H to the lower right.

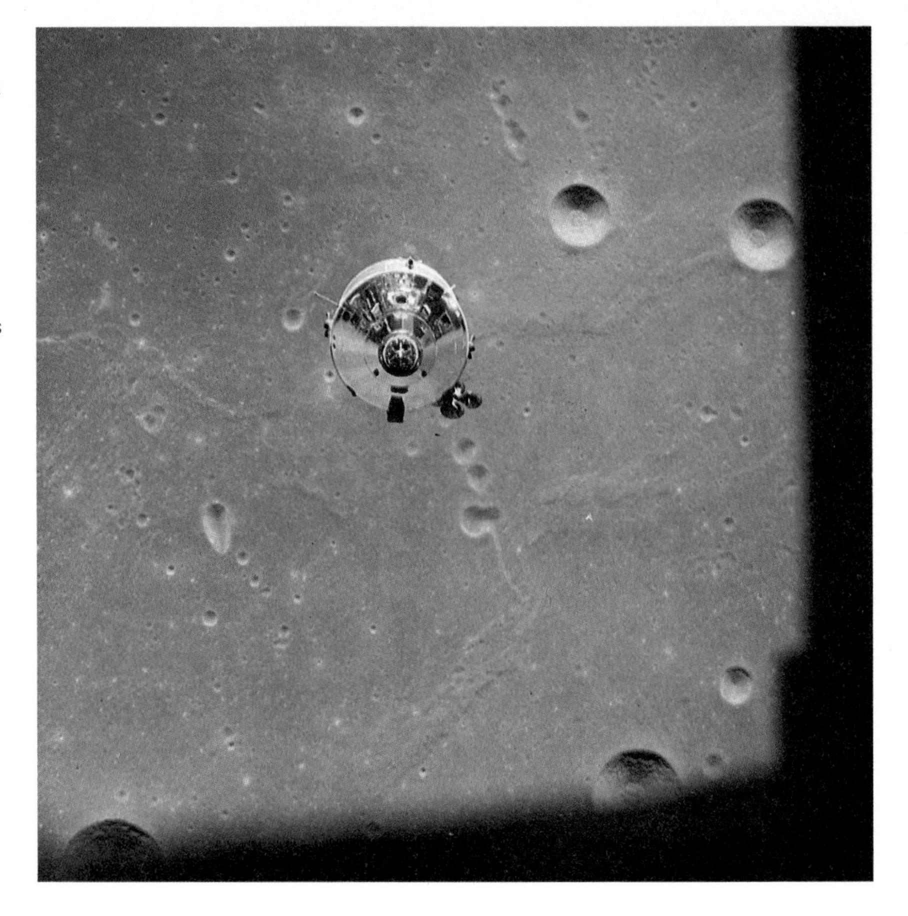

the surface about two inches. Although the surface appears to be very finely grained as you get close to it, it's almost like a powder. Now and then it's very fine. I'm going to step off the LM now. That's one small step for a man, one giant leap for mankind.'

A second later, at 2.56 G.M.T., Neil Armstrong was standing on the Moon.

An untrained man would have wasted precious time in gazing around. Armstrong did not. He could stay outside for less than three hours – his oxygen supply was limited – and there was much to do. At once he used his scoop to collect some material from the lunar surface, in case of the need for a sudden return to the Module, but the ground seemed to be pleasingly solid, and there was no difficulty in walking, though the low gravity, only one-sixth as strong as on Earth, meant that everything seemed to happen in slow motion. Aldrin joined him, and the two worked methodically, setting up scientific equipment and taking photographs. The experiments were of various kinds; for instance there was a seismometer, designed to record any 'moonquakes', and also a strip to catch particles of solar wind (dismantled at the end of the moon-walk, and brought back). But the most important requirement was the collection of lunar samples.

Apart from a brief pause to talk with President Richard Nixon, there was no respite. As they worked, they kept up a running commentary.

OPPOSITE: Apollo 11; Colonel Edwin Aldrin on the Moon, setting up the solar wind experiment with the Lunar Module in the background. (Most of the pictures of the Apollo 11 moon-walk show Aldrin, because Neil Armstrong was taking the photographs.)

*Intrepid*, the lunar module of Apollo 12, standing on the Moon; Astronaut Charles ('Pete') Conrad beside it. The rocks brought back from this site differ slightly from those obtained from Apollo 11, but in general the samples returned from the six successful Apollo missions, as well as the Russian automatic probes, are of the same type.

'You have to be rather careful to keep track of where your centre of mass is,' said Aldrin. 'Sometimes it takes two or three paces to make sure you've got your feet underneath you.' There was something eerie about it all, and yet the two astronauts were so very matter-of-fact and practical. 'Magnificent desolation,' was how Aldrin described the lunar scene.

The sample collection was completed; the experiments were checked; then first Aldrin, then Armstrong went back inside the *Eagle*. The first moon-walk was over. Armstrong had been outside for 2 hours 47 minutes, and Aldrin almost as long.

Following a somewhat fitful sleep, preparations were made for blast-off. Now everything depended upon *Eagle*'s single ascent engine, but it worked perfectly; using the bottom part of the Module as a launching pad the top portion was fired into orbit, and made rendezvous with the waiting Collins. Docking was accomplished without much trouble, after which the *Eagle* was jettisoned, and the return journey began. At 4.40 p.m. on 24 July the Command Module carrying the three astronauts splashed down in the Pacific. After a total of 195 hours away from Earth, the final landing was a mere 30 seconds late.

This time, because the astronauts had been in contact with lunar material, the reception was different. The chances of their having brought back anything harmful were remote, but not nil, so that strict quarantining was imposed, and it was not until every possible test had been made that the three were pronounced 'clear'. By then the precious samples had been studied, and the first records had come back from the seismometer that had been left at their moon landing site, Tranquillity Base.

A few months later Neil Armstrong joined me for a 'Sky at Night' programme, and when I asked him about the possibilities of a fully-fledged Lunar Base his reply was quite positive. 'I'm certain we'll have such a base in our lifetimes. Somewhat like the Antarctic stations and similar scientific outposts; continually manned, though there's certainly the problem with the vacuum and the high and low temperatures of day and night. Still, in some ways it's more hospitable than the Antarctic. There are no storms, no snow, no high winds, no unpredictable weather; as for the gravity – well, the Moon's a very pleasant kind of place to work in. It's quite practicable.'

I wonder when this will really happen? It is not in the least far-fetched, and Neil's optimism was shared by Wernher von Braun, who was quite confident that the first Lunar Base would be set up before the year 2000. (Von Braun had been at Mission Control throughout the flight of Apollo 11; it must have been a tremendously satisfying time for him.) However, it is important to remember that magnificent though it was, Apollo 11 was purely a reconnaissance mission. Before any permanent base can be established, rockets much more powerful and, incidentally, much cheaper will have to be developed.

Apollo 12, carrying Astronauts Conrad, Bean and Gordon, went up on 14 November 1969. This time the target point was in the Oceanus Procellarum, near the automatic soft-lander Surveyor 3, which had been on the Moon ever since April 1967 and had sent back well over 6000 pictures before its power failed. It was hoped to land so close to Surveyor that the moon-walkers, Conrad and Bean, would be able to go across to it and bring back sections of it for analysis. I remember voicing my doubts as to whether the landing would be accurate enough for this, but I was wrong; Surveyor was reached, and the whole mission was successful. It was with the next expedition, the ill-fated Apollo 13, that trouble came.

At first all seemed well. James Lovell, Fred Haise and Jack Swigert were on their way, 178,000 miles from Earth, when disaster struck. Lovell's voice came through: 'Hey, Houston, we've got a problem here' – not only the understatement of the year, but the understatement of the century. There had been a violent explosion in the Service Module. Part of the hull was blown away, and the power supply was cut off instantly, so that Apollo 13 was well and truly crippled.

Nobody who followed the course of events over the next few days is likely to forget them. Improvisation was the only hope. The remaining power supplies had to be husbanded; even the lights were turned out, and as the spacecraft drifted on the cabin temperature dropped alarmingly. There was a potential shortage of water, and – worse – oxygen. The Service Module, an essential part of Apollo, had become a useless hulk.

A spacecraft cannot be turned round at short notice. The only chance left was to continue round the Moon, and then use the descent engine of the Lunar Module to accelerate the whole vehicle and bring it back to Earth at a suitable splashdown point. Moreover the fastest return orbit would be the most dangerous, since it would leave no power reserves at all. A compromise was worked out at Houston; once Apollo 13 had passed behind the Moon, the Lunar Module engine was fired, and after two more course corrections Apollo 13 was back in the neighbourhood of the Earth. It was with heartfelt relief that the millions of television viewers saw the Command Module splash down safely.

The cause of the trouble was that a switch in the main electrical system

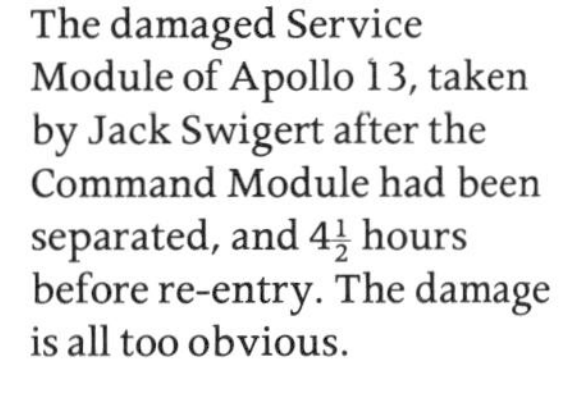

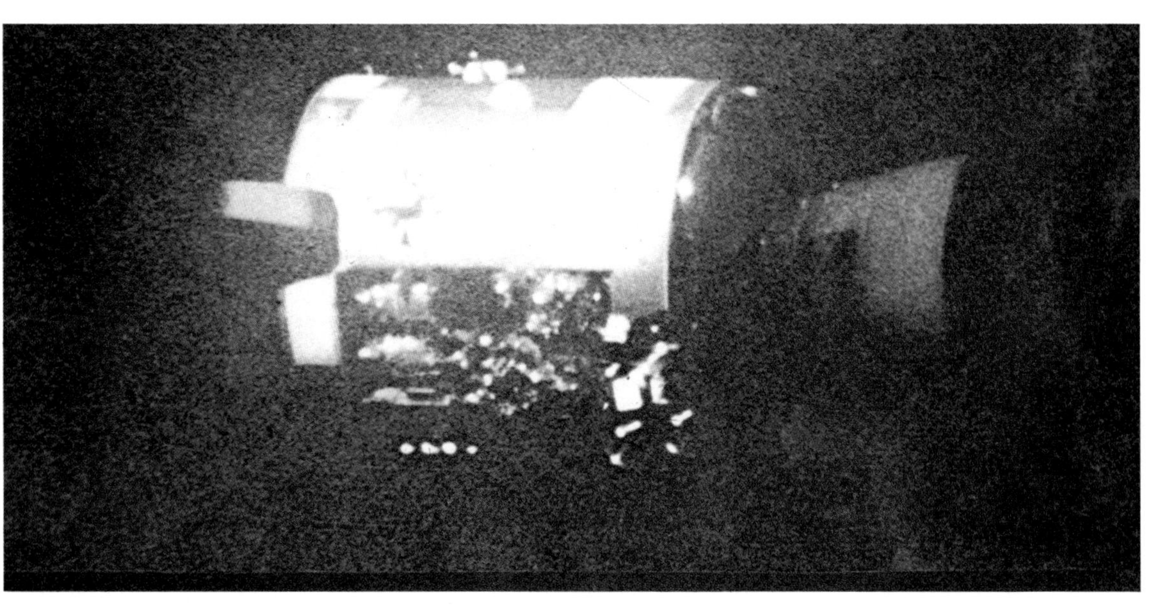

The damaged Service Module of Apollo 13, taken by Jack Swigert after the Command Module had been separated, and $4\frac{1}{2}$ hours before re-entry. The damage is all too obvious.

Apollo 15, showing the Lunar Module and also the LRV or Lunar Roving Vehicle in which David Scott and James Irwin drove around the Moon. In the background is one of the Apennine peaks. There is nothing jagged about it; David Scott described it to me as 'a featureless mountain'.

had been welded shut instead of being left free. Temperatures rose, leading to overheating and then a fire. Had the problem arisen on the return journey instead of the outward one nothing could have been done, because the Lunar Module would have been jettisoned after the landing on the Moon. At least Apollo 13 showed that the spacecraft controllers were flexible, but it had been a desperately close shave, and lesser men than Lovell, Haise and Swigert would have had no chance of survival whatever Houston might have done.

One point that struck me forcibly was that during the crisis all national differences seemed to be temporarily put aside, so that for once the world was united. Scientifically, Apollo 13 was a failure; from the human viewpoint it was a tremendous triumph.

I will pass more briefly over the next three Apollos, important though they were. In February 1971, Apollo 14 came down in the Oceanus Procellarum near the walled plain Fra Mauro, carrying Alan Shepard together with Astronauts Mitchell and Roosa; this time the two moon-walkers took a specially designed cart with them, so that they could extend their range and spread their equipment more widely. Apollo 15 (Astronauts Scott, Irwin and Worden) landed in the foothills of the Apennine Mountains, and using an electrically powered moon-car Scott and Irwin drove to the very edge of the deep crack known as the Hadley Rill. Apollo 16 landed in the highlands in the Moon's southern region, taking Astronauts Young and Duke to the surface. Finally, on

Apollo 17: Astronaut Ronald Evans outside the spacecraft during the return journey to Earth. During this 'spacewalk', Evans retrieved film cassettes from the various instruments which had been used during the time spent in lunar orbit; the cylindrical object to his left is the Mapping Camera cassette. Evans was outside the spacecraft for 1 hour $7\frac{1}{4}$ minutes; the date was 17 December 1972.

7 December 1972, came Apollo 17, which was scientifically the most valuable trip of all.

The commander was Eugene Cernan, a veteran of Apollo 10. With him was Dr Harrison ('Jack') Schmitt, a professional geologist, who had been given astronaut training specially for the occasion; the third member of the team, Astronaut Evans, orbited the Moon during the landing expedition. The target area, Taurus-Littrow on the edge of the dark plain of the Mare Serenitatis (Sea of Serenity), had been chosen because it was expected to be of particular interest geologically, with ancient rocks dating back over at least 4000 million years. Again a moon-car was taken to the surface, and the equipment included several new experiments. One of these was designed to measure the heat coming from inside the Moon, which involved drilling a hole and thrusting recording instruments inside. This experiment had failed with Apollo 16, because a vital cable had been wrenched loose when Duke tripped over it, but careful re-designing had ensured that nothing of the kind could happen again, and the results were very significant. There had been considerable discussion as to whether or not the Moon had a hot core. The Apollo 17 measurements showed that it has, although the internal temperature is much lower than that of the Earth.

I was at Cape Canaveral for the launch – the only one I actually saw except on television – and it was a fascinating experience. Blast-off was postponed several times because of minor technical problems, and finally took place after dark, which made it even more dramatic. As the engines were ignited I could see smoke pouring from the exhaust; even at a distance of four miles (the nearest one could safely get) the brilliance was dazzling, and as the rocket lifted off there was what I can only describe as a deafening 'wave of sound'. It was hard to credit that three men were right in the middle of what looked like an inferno.

When I reached Houston I found that the atmosphere inside Mission Control was tense but confident. Communications were excellent, the pictures from the lunar surface were clear, and it was plain that Schmitt's expert geological knowledge was invaluable. During the second moon-walk of this expedition, on 12 December, there was added excitement. Near a craterlet which had been nicknamed Shorty, Schmitt suddenly found coloured material, and his words came through: 'It's orange! Crazy!' Nothing of the sort had been expected, and most people (including myself) thought that it might indicate recent volcanic activity, but later analysis showed that this was not so, and that the colour was due to numerous small, glassy beads that were very ancient indeed.

The landscape was of the usual type, with huge boulders which sometimes hid the astronauts from view. Another craterlet some way from Shorty looked like a field of rocks, and gave every impression of having been formed by a meteoritic impact rather than by volcanic collapse. Finally Cernan and Schmitt re-entered the Lunar Module, their work done. When they blasted away, direct exploration of the Moon was temporarily at an end.

Almost ten years later I had a long talk with Commander Cernan in Houston, where he is now a highly successful businessman. Obviously his memory of the trip was as vivid as if it had happened only yesterday, and he told me that he had been particularly fascinated by the Earth – 'beautiful and majestic; it was hard to realize that I was looking at it through black space filled with sunlight'. When the Apollo was in orbit

Astronaut Harrison ('Jack') Schmitt, the only professional scientist to have been to the Moon, standing by a huge boulder. Dr Schmitt replaced Colonel Engle on the Apollo 17 mission because he was a trained geologist. Colonel Engle subsequently piloted the second flight of the Space Shuttle in 1981.

round the Moon and passed into shadow, it was 'the blackest black that a human being can imagine. You're out of sight of the Earth, out of sight of everything except the stars; you can't even see the surface of the Moon below'.

Lunar dust, he said, was probably the most difficult problem that had to be faced. 'It's like graphite; but graphite lubricates, while lunar dust tends to make everything stick together. It gets into spacesuits and all moving parts, and when you get back into the Module and unpack the rocks, holding them in your hand and feeling them, the dust is so fine that it penetrates into your skin. It was weeks after getting back before I could get it out.'

Navigation on the lunar surface was not hard, because even though distances were difficult to estimate accurately the Lunar Module could be kept in view most of the time. 'We landed in a beautiful valley surrounded by mountains, and we carried out a lot of geological work, though I believe it may be 50 or 100 years before we know which were the most important results.' The orange soil was very much of a surprise; 'When Jack Schmitt first yelled out, I almost reckoned that he'd been on the Moon one day too long!'

Finally I asked Gene Cernan whether he would care to make a return journey, and he was quite definite. 'I'd go back tomorrow if I could. I'd like to go back and take time to recapture those moments in my life which went by so quickly. There were so many things to do, and which your life depended on your doing, that there was no chance to stand back and enjoy it all.'

Originally it had been planned to continue the Apollo series up to No. 21, but it was certainly wise to call a halt with 17. Had the missions continued, sooner or later something would have gone badly wrong, and moreover Apollo had done everything that it could reasonably be expected to do. Neither must we forget the Russians, who have scored major successes. They have sent two unmanned probes to the Moon and

Commander Eugene Cernan, the last man on the Moon, photographed here with me during a television film recording in November 1981. Photograph by Kevin Baxendale

Lunokhod 1; a Russian representation of the probe standing on the Moon's surface. The vehicle was taken to the Moon by the probe Luna 17; the landing took place on 17 November 1970 in the Mare Imbrium, not far from the beautiful Sinus Iridum. During the next 11 months Lunokhod travelled a total distance of 16.8 miles, sending back over 200 panoramic pictures and 20,000 photographs. The second Lunokhod was taken to the Le Monnier region, near the edge of the Mare Serenitatis in January 1973, and before its power failed on the following 3 June it had sent back 86 panoramic pictures and 80,000 photographs, during which time it travelled a distance of almost 60 miles. Both Lunokhods are still standing on the Moon, and their precise positions are known, so that they will no doubt be recovered eventually.

brought them back together with samples, and they have also landed two 'crawlers' or Lunokhods, which looked like demented taxicabs but which proved to be extremely effective. Like the abandoned Apollo moon-cars and the grounded rockets, they also will remain *in situ* until someone goes to fetch them.

All in all, what have the space missions told us about the Moon as a world?

Our knowledge has been increased vastly since those early flights. We know now that there is a loose outer layer or regolith, which has been continually churned by the impacts of tiny meteoritic particles, and which in places has a depth of almost 100 feet. The thickness of the crust is generally between 30 and 40 miles, while the mantle underneath goes down to about 600 miles. Below this again is a region known as the asthenosphere, which may be partially melted, and finally there is the molten core, rich in iron and at a temperature of perhaps 1500 °C. All the Moon's rocks are volcanic. There are no substances unknown on Earth, though some of the rocks differ in structure because they solidified under very different conditions.

Tremors in the globe do occur; moonquakes may be either shallow or deep seated, but they are very mild, and will pose no dangers to future Lunar Bases. The rocks contain no trace of hydrated materials (that is to say, water in any form), and there is absolutely no sign of life, either past or present. The Moon has been sterile throughout its long history.

At present the Moon is once more being somewhat neglected. The equipment left behind by the Apollo astronauts has been switched off, and no messages are being transmitted, while the main attention of the space planners has shifted elsewhere. I have no doubt that things will change again before long, and that a Lunar Base will be in existence in the foreseeable future. But whatever may happen, the Moon remains a world of special interest to us. Despite the Lunas, the Orbiters, the Surveyors and the Apollos, it has lost none of its magic.

# 4 The Mariners

Cape Canaveral, Mission Control at Houston, and other centres of space technology are all impressive in their different ways. To me, however, perhaps the most striking of all is a quiet room not far from Pasadena, outside Los Angeles. This is the Deep Space Network or D.S.N., in the heart of the Jet Propulsion Laboratory.

Visitors can see the room only from behind a glass screen. When I went right inside for the first time I had a momentary feeling of having been transported to some future age, and I felt just the same when I went there a few weeks before writing this chapter. Not that there is anything outwardly spectacular; there are the usual control desks and panels, and subdued lighting, while one of the walls is covered by screens that list data and show images of Jupiter, Saturn and other planets. Yet it is from here that we communicate with our furthest outstations, the planetary probes.

The Jet Propulsion Laboratory or J.P.L. was founded in 1958 by Theodore von Kármán. The first successful planetary spacecraft, Mariner 2 to Venus, was launched four years later, in August 1962, and in the following December it passed within 22,000 miles of its target, sending back the first reliable information about that decidedly peculiar world. Mariner 2 was controlled from Pasadena, and ever since then the D.S.N. has been manned 24 hours a day. There has been talk of moving J.P.L. to Cape Canaveral, but whether this will ever happen I do not know.

Equipment is needed not only at Pasadena, but at suitably distributed stations throughout the world. In the D.S.N. there are three main stations. One is at Goldstone in California, one at Canberra in Australia, and the third at Madrid in Spain. (The Madrid station was transferred there from Hartebeespoort in South Africa some years ago, rather unwisely in view of the fact that the weather in Spain is less reliable, and heavy rain at the wrong moment can cause trouble.) Each station has a radio 'dish' 210 feet in diameter, together with two smaller radio telescopes. The fact that the three stations are so widely separated means that a probe must always be above the horizon from at least one of them, so that contact can never be lost. Together with the radio telescopes, each station is provided with transmitting, receiving, data handling and interstation communication equipment.

The D.S.N. room at the Jet Propulsion Laboratory is the nerve-centre of the whole operation. All the data received are sent there. When a command has to be sent to a distant probe, it is passed by the Mission

Venus, from the Pioneer orbiter: 14 February 1979. At this time the Pioneer was 36,580 miles from the planet. This image looks unusual because the terminator is at the bottom of the picture. The prominent features on this image take the forms of belts around the equator; the bright polar cloud in the southern hemisphere shows a disturbance extending from near the terminator towards the equator. The arms of a familiar cloud structure, known as the Y, can be seen appearing over the limb.

Inside the mission operations control room at the Johnson Space Flight Center at Houston, Texas. Terry J. Hart is at the centre, and Daniel C. Brandenstein to his right. At this moment John Young and Robert Crippen were in the cabin of the Space Shuttle orbiter *Columbia* on the launch pad at Complex 39, Cape Canaveral, awaiting the moment of launch.

Controller to the best-placed outstation, where it is checked for accuracy and then transmitted.

Looking round the quiet room, it is hard to appreciate just how much technology is involved. Something may go wrong with a spacecraft more than 1000 million miles away – a jammed camera, perhaps, or a fault in the direction-finding sensors; what can be done? This is for the Mission Controller to decide. Having made up his mind, he sends out his orders. Hours later they reach the probe, and more often than not they are obeyed. The initial responsibility lies at Cape Canaveral, but as soon as a spacecraft has been launched, and is clear of its gantry, the D.S.N. takes over.

The amount of power received from a planetary probe is incredibly small, and not nearly enough to light up the bulb of a pocket torch for a single second. Yet from this we have been able to receive not only the pictures, but also scientific data of many kinds. And at any time the D.S.N. operators will know the exact distance, speed, attitude and operating status of any probe with which we are still able to communicate, even though some of them may have been in space for many years.

Sending a spacecraft to a planet is much more difficult than to the Moon. Not only are the distances greater, but the planets orbit the Sun rather than the Earth, so that they do not stay conveniently within range. Venus, the closest of them, is always at least 100 times as remote as the Moon, so that a journey there takes months instead of only a few days. Moreover, there is no chance of carrying out running repairs.

Everything depends on the commands sent out from the D.S.N.

Things are made even more troublesome because it is impossible to go by the shortest route, since this would mean using rocket power throughout the flight, and no probe could carry enough fuel. Also, absolute accuracy is essential, and though the launcher must attain a peak velocity of seven miles per second the probe itself must have its speed controlled to within a few miles per hour. If not, it will miss its target by a wide margin.

In the case of Venus, which is closer to the Sun than we are, the procedure is to take the probe up by the usual step-rocket arrangement, and then 'slow it down' until, relative to the Sun, it is travelling more slowly than the Earth. It will then start to swing in towards the Sun, and if all goes well it will reach the orbit of Venus at the correct moment and position for an encounter. Most of the journey is done in free fall, so that the probe coasts towards its target without having to use any of its own power.

The first such attempt was made by the Russians in February 1961. Unfortunately, contact with their spacecraft, Venera 1, was lost within a few weeks; at that time the Soviet workers were having great problems with long-range communication (problems which, at the time of writing, they have not completely solved). Mariner 1, the first American probe, was also a failure, and did not even achieve an Earth orbit. But Mariner 2, sent up on 26 August 1962, was very different, and provided much information, some of which was unexpected and even rather unwelcome.

The lower stage of the step-launcher was a powerful rocket called an Atlas. It rose vertically from Canaveral, and then moved off in the general direction of South Africa. The second stage, an Agena rocket,

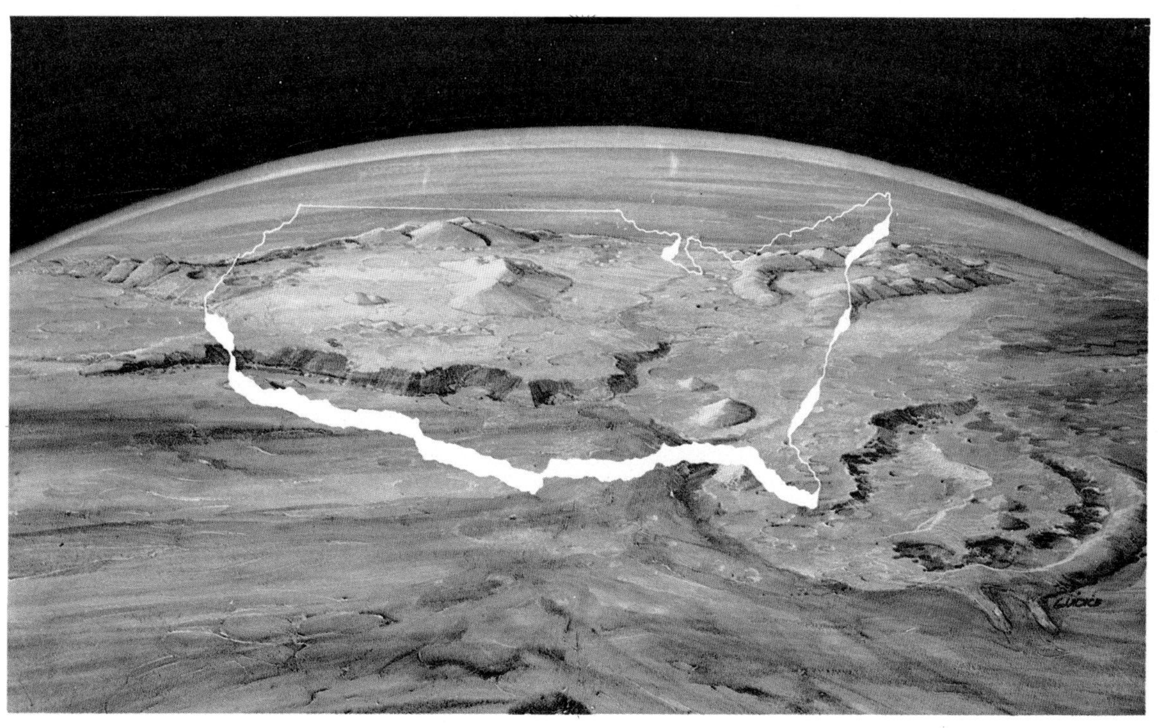

Aphrodite Terra, on Venus. This is an artist's impression, based upon results from the Pioneer orbiter. An outline of the continental United States has been added to provide a scale. Aphrodite, 6000 miles long by 2000 miles wide, lies almost parallel with the equator of Venus, centred at latitude 5 °S., longitude 100 °E. It is made up of eastern and western mountain areas, separated by a lower region; the western mountains rise to 26,000 feet above the mean level of Venus. The separating lowland region includes the great rift valley.

Artist's impression of the surface of Venus, based upon results obtained from the Pioneer orbiter; these measurements cover 93 per cent of the total surface, extending from latitude 75 °N. to 63 °S. It must be borne in mind that this impression is of the surface features, which can never be actually seen from above the top of the obscuring atmosphere of the planet.

then took over; the Atlas fell away, and the Agena-Mariner combination was put into an Earth orbit at a height of 115 miles, moving at the regulation 18,000 m.p.h. or 5 miles per second. (This, of course, is the velocity that a probe must reach if it is not to fall back to the ground.) As it reached the African coast, the Agena was fired again, and the probe's total velocity increased to 25,503 m.p.h., which is 850 m.p.h. more than the escape velocity at that height.

As the Mariner moved away, it was slowed by the Earth's pull of gravity, until at a distance of 600,000 miles its velocity relative to the Earth had fallen to only 6874 m.p.h. In other words, Mariner was now moving round the Sun at a velocity of 6874 m.p.h. less than that of the Earth, and so it began to swing inwards, picking up speed as it went. Meanwhile the Agena rocket had been separated from the Mariner, rotated through a wide angle and then fired again, so that it was put into an entirely different path. Nobody knows what happened to it, and nobody cares; it had completed its task, and on the long journey to Venus it would have been useless. Henceforth, Mariner 2 was on its own.

This was only the beginning. Mariner was in the right orbit, but it had to be kept there; it also had to be kept pointing in the correct direction, as otherwise it would have been unable to communicate with Earth. Special sensors were used to lock on to the Earth and the Sun (many later probes have used the Sun and a bright southern star, Canopus). A midcourse correction was made on 14 December, but even though the final minimum distance from Venus was 21,594 miles instead of the planned 10,000, Mariner 2 was able to carry out its full programme. It travelled past Venus, sent back its data, and then continued on its way. Contact was finally lost on 4 January 1963, by which time it was 54,000,000 miles from the Earth. No doubt it is still moving round the Sun; unless it is hit and destroyed by some solid body, it will remain in orbit indefinitely. It has become a permanent member of the Solar System.

Up to the time of the Mariner 2 encounter, Venus had been generally described as 'the planet of mystery'. We knew surprisingly little about it, which was why the Mariner revelations were awaited with such keen interest.

Because of its nearness, and because of its size, Venus shines as the most brilliant object in the sky apart from the Sun and the Moon. When at its best, in the west after sunset or in the east before dawn, it can sometimes cast a perceptible shadow, and really keen-sighted people can see it with the naked eye even when the Sun is well above the horizon. It is only slightly smaller than the Earth – according to the latest measurements its diameter is 7519 miles, compared with our 7926 – and it is highly reflective, because it is covered by a dense, cloud-laden atmosphere.

The trouble is that even when a large telescope is used, practically no detail at all can be seen! I have been observing Venus since 1934, and I have never been able to make out more than a few hazy, ill-defined and temporary patches which are undoubtedly atmospheric. Of course, Venus like the Moon, shows phases or apparent changes of shape from new to full, because we see different amounts of its sunlit hemisphere, but to all intents and purposes the surface looks blank.

It was easy enough to measure the length of Venus' 'year' (the time taken for one orbit of the Sun): it was found to amount to $224\frac{3}{4}$ Earth-days. Venus is not only closer to the Sun than we are (67,000,000 miles on average) but is also moving more quickly. However, the length of the Venus 'day' was unknown, and estimates ranged between 24 hours and well over a month.

Before 1965, the one positive piece of information was that the atmosphere of Venus is not the same as ours. The Earth's air is made up chiefly of two gases, nitrogen and oxygen. In the atmosphere of Venus, the most plentiful gas was found to be carbon dioxide. This had been discovered by means of the spectroscope, an instrument that splits up light and shows us the chemical composition of the light-source. Unfortunately there was no way of seeing below the top of the cloud-layer, and so what the surface conditions were like seemed to be anybody's guess.

Before the Mariner flight, there were two main theories. Venus might be a raging dust-desert, with a fiercely hot surface. Or the whole planet might be covered with water, with only a few islands poking out here and there. The second theory, due to two eminent American astronomers (D. H. Menzel and F. L. Whipple) was certainly the more attractive, and it indicated that there might be some sort of life on Venus. Thousands of millions of years ago the Earth's air contained much more carbon dioxide and much less free oxygen than it does now, and it was then that life first appeared in the oceans. Could the same sort of thing be happening today on Venus? It seemed at least possible. (*En passant*, the atmospheric carbon dioxide would have fouled the oceans, producing seas of soda-water – though not even the greatest optimist would expect to find any whisky to mix with it.)

Another puzzle concerned the so-called Ashen Light. When the Moon is a crescent, the 'dark' side can often be seen shining faintly; this is the appearance often called 'the Old Moon in the New Moon's arms', and is easily explained as being due to the light reflected from the Earth. The same sort of thing has been reported many times with the crescent Venus, and I have observed it so strongly that I am convinced that it cannot be due to a contrast effect or any other type of illusion. It is not easy to account for, because Venus has no satellite (and I cannot agree with the last-century astronomer Franz von Paula Gruithuisen, who believed it to be caused by vast forest fires lit on the planet's surface by the local inhabitants to celebrate the election of a new Government!). It is probably due to electrical activity in Venus' atmosphere, similar to our auroræ or polar lights but much more brilliant.

Such was the state of our knowledge – or lack of it – when Mariner began its journey. By the end of December 1962, many of the most important problems had been cleared up.

Mariner 2 carried instruments to measure the surface temperature. The results were absolutely conclusive: Venus was very hot indeed, and a thermometer on the surface would register around 900 °F., so that the Menzel-Whipple marine theory had to be abandoned. No water could possibly exist in such conditions, and the idea of life on Venus was virtually ruled out.

There was also the question of the surface pressure. Venus' atmosphere is much denser and more extensive than ours. At the surface, the pressure is between 90 and 100 times that at sea-level on Earth, so that a visitor would certainly be squashed as well as cooked and suffocated.

Another revelation was that Venus takes 243 days to spin once round, so that technically its 'day' is longer than its 'year', although from any point on the surface the interval between one sunrise and the next would be only 118 days. Why Venus has this slow spin is a mystery, and to complete the picture its rotation is from east to west instead of west to east. Finally, Mariner 2 reported that there was no trace of any magnetic field.

When these results came through, most people were taken aback. Why should Venus spin in what might be called an upside-down direction, and why was the surface temperature so high?

Sunrise on Venus, as seen from the Pioneer space probe in orbit round the planet. The photograph was taken on 5 December 1978, soon after Pioneer was put into a closed path making one full circuit round Venus every 24 hours. No detail is visible; all that can be seen is the top of the cloud-layer.

Mariner 2 could tell us no more, and the next attempts came from the Soviet Union. Let it be said at once that since their initial failure the Russians have had great success in probing Venus, even though they have consistently failed with Mars, which should be a much easier target.

In 1965 two spacecraft were sent up. The second of them (Venera 3 in the whole series) was ambitious, because it was designed to make a parachute descent through the atmosphere and land gently enough to transmit data from the surface. This was a particularly difficult task

because there could be no last-minute corrections; everything had to be purely automatic, and in view of the high temperature the spacecraft had to be chilled before plunging into the dense clouds. No data were received, though until the start of the final manœuvre everything had seemed to be going well. In the following January there were two more attempts, with Veneras 5 and 6, and this time some data were transmitted during the descent, though the results did not seem to fit in with the information sent back by Mariner 2 or by Mariner 5, an American probe that had flown past Venus in 1967 at a minimum distance of only 4000 miles.

Eventually the answer was found. The lower atmosphere of Venus was not only hotter than had been expected, but also denser. The Russian probes had been literally crushed well before landing, so that the data sent back referred not to the surface, but to a region well above it.

The next Veneras, numbers 7 and 8, were suitably toughened, and landed successfully, transmitting data for some time after arrival. They could not survive for long under those inferno-like conditions, and within an hour they 'went silent', but they had had time enough to provide full confirmation of what the surface was now believed to be like.

A new complication was becoming evident. There had always been doubts about the composition of Venus' clouds, and all sorts of compounds had been suggested, but the spacecraft showed that there was a large amount of sulphuric acid, which is an unpleasant substance and is highly corrosive. The more one learned about Venus, the more hostile it appeared to be.

Passing over Mariner 10, which by-passed Venus in February 1974 on its way to the inner planet Mercury, we come to Veneras 9 and 10, sent up by the Russians in 1975 and which landed within a few days of each other – Venera 9 on 21 October, and its follower on 25 October. Each was able to transmit one picture before being put out of action, and it was seen that the landscape was gloomy by any standards. The Venera 9 site was described as being like 'a heap of stones', while the Venera 10 panorama was smoother, and looked as though it might be an old plateau. The light-level was rather brighter than had been feared, and was compared with that in Moscow at noon on a cloudy winter day. The probes had taken searchlights with them in case they had been plunged into inky blackness, but there was no need to use them.

What about the rest of the surface? This was where radar came into the reckoning, first with Earth-based equipment and then from a Pioneer

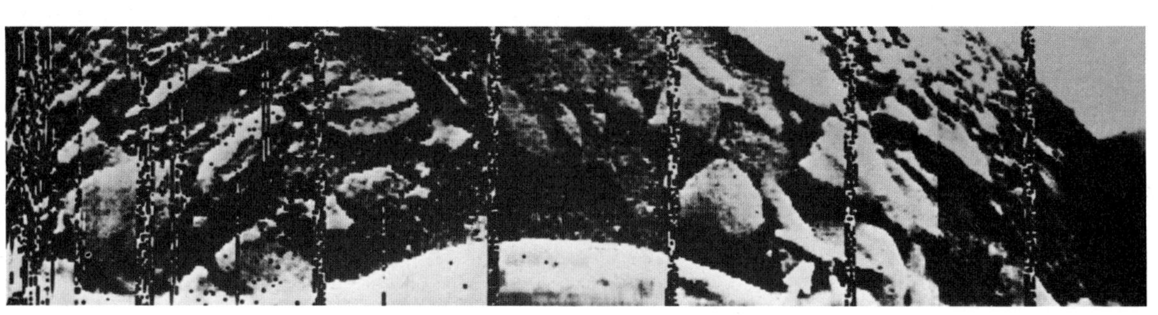

The first photograph obtained direct from the surface of Venus, by Venera 9 on 21 October 1975. The probe was able to transmit for 53 minutes after arrival. The light curved area in the foreground is part of the landing gear; to the right (indicated by the arrow) is a piece of equipment lying on the surface. The vertical stripes indicate the sections in which information about the work of the scientific apparatus was transmitted. This photograph is not touched up.

probe launched from Canaveral in May 1978, which reached Venus in the following December and was put into orbit round the planet.

Radar consists, essentially, of sending out a pulse of energy and then receiving the 'echo'; it has been compared, admittedly rather loosely, with throwing a tennis-ball against a wall and catching it on the rebound. The nature of the echo gives information about the object off which it has been bounced, and there had already been indications that Venus was a world containing large, rather shallow craters which were presumably volcanic. The radar equipment on Pioneer was much more informative, and we now have good maps of most of the surface of Venus even though we cannot see it directly. Pioneer was not the only spacecraft to reach Venus in December 1978; there were two Soviet Veneras and also a second Pioneer which dispatched several mini-probes to the surface, but it was the radar results that were so significant.

The maps cover almost all the region between latitudes 75° N. and 63° S. There are highlands, lowlands, and a vast rolling plain that extends well over half the surface. There is no ocean, so that we cannot take sea-level as a standard, but if we reckon from the mean radius of the planet we find that the total range of altitudes is from 35,000 feet above the mean level to 9500 feet below. This means that the highest mountains on Venus are loftier than Everest.

OVERLEAF: Map of Venus, compiled from the Pioneer Orbiter results. The highlands are shown in green, yellow and red; lowlands in dark blue; and the huge rolling plane, covering about 60 per cent of the surface and varying in height only by about 3300 feet, in light blue and blue-green. Ishtar Terra, in the northern hemisphere, is shown by yellow and brown contours; it is about the size of Australia. The Maxwell Mountains, to the right (east) side of Ishtar are the highest peaks on the planet. Aphrodite Terra, near the equator, is larger than Ishtar, but appears smaller because this map is drawn on Mercator's projection – just as a Mercator map of the Earth will show Greenland as larger than Africa. The two highly volcanic areas, Beta Regio and the Scorpion's Tail of Aphrodite (around latitude 0°, longitude 200°) are well shown. Note that most of the names allotted to the various features by the International Astronomical Union are female – this was regarded as appropriate for a planet named after the Goddess of Beauty!

There are two major uplands, which have been named Ishtar Terra and Aphrodite Terra. Ishtar, in the north, is the smaller of the two, about the size of Australia. To its east are the Maxwell Mountains, the highest peaks on the planet, topped by a dark circular feature 3000 feet deep, probably a volcanic caldera. Closer to the equator lies Aphrodite Terra, roughly half the size of Africa, made up of two mountainous areas separated by a lower region; like Ishtar, it is bounded by steep scarps, and at its eastern end the huge Rift Valley: 175 miles wide at its broadest point, 1400 miles long, and well over 9000 feet deep. Presumably it has been caused by a tremendous break in the planet's crust. Far over to the west are Rhea and Theia, two shield volcanoes built up by material sent out from below the ground. There are shield volcanoes on Earth – Mauna Loa in Hawaii is one – and for that matter on Mars too; but are Rhea and Theia active or extinct? Very recent results indicate that they are active.

On Earth, the crust is broken into several continent-bearing 'plates', which drift very slowly around as though floating on the molten interior; the science of what is now called plate tectonics has become very important indeed. But Venus has a thicker crust, and is a one-plate planet, so that the entire situation is different. The Earth's interior heat is vented at many points, but with Venus there are apparently two main regions – Beta Regio, made up of Rhea and Theia, and an area known as the Scorpion's Tail of Aphrodite Terra. It is here that we find the volcanoes, though no doubt there are others as well.

There is another feature of this one-plate situation. Crustal density measurements indicate that there has been local uplifting of large regions, probably because of up-flowing convective 'plumes' resulting from the circulation of the interior magma. This accounts for the presence of Aphrodite and Ishtar. Vertical movements in the crust are indicated by the deep rift valleys, one at the lowest point on the planet – near the

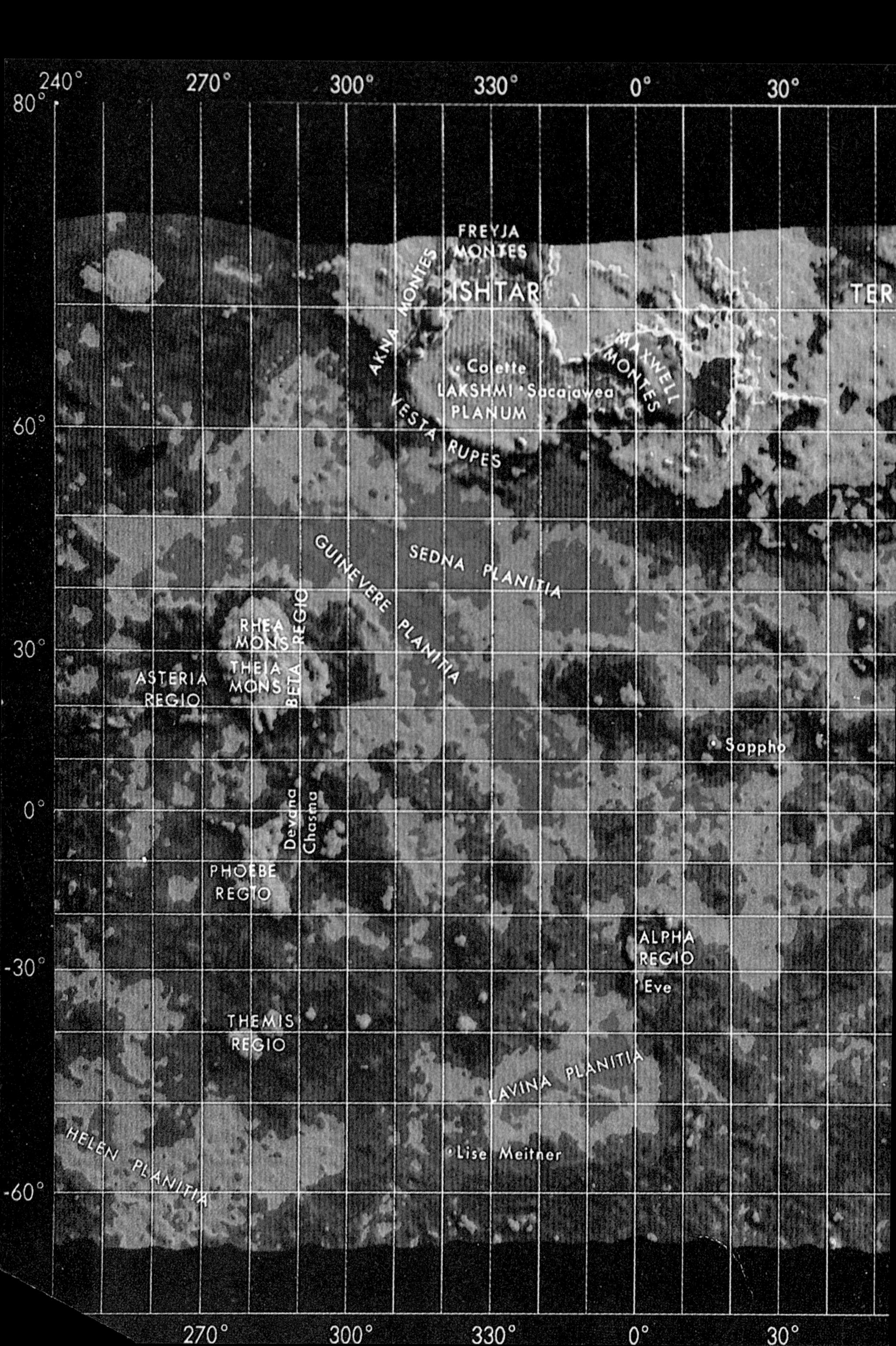

240°
270°
300°
330°
0°
30°
80°
60°
30°
0°
-30°
-60°
FREYJA MONTES
ISHTAR
TER
AKNA MONTES
MAXWELL MONTES
Colette
LAKSHMI PLANUM
Sacajawea
VESTA RUPES
SEDNA PLANITIA
GUINEVERE PLANITIA
RHEA MONS
THEIA MONS
BETA REGIO
ASTERIA REGIO
Sappho
Devana Chasma
PHOEBE REGIO
ALPHA REGIO
Eve
THEMIS REGIO
LAVINA PLANITIA
HELEN PLANITIA
Lise Meitner
270°
300°
330°
0°
30°

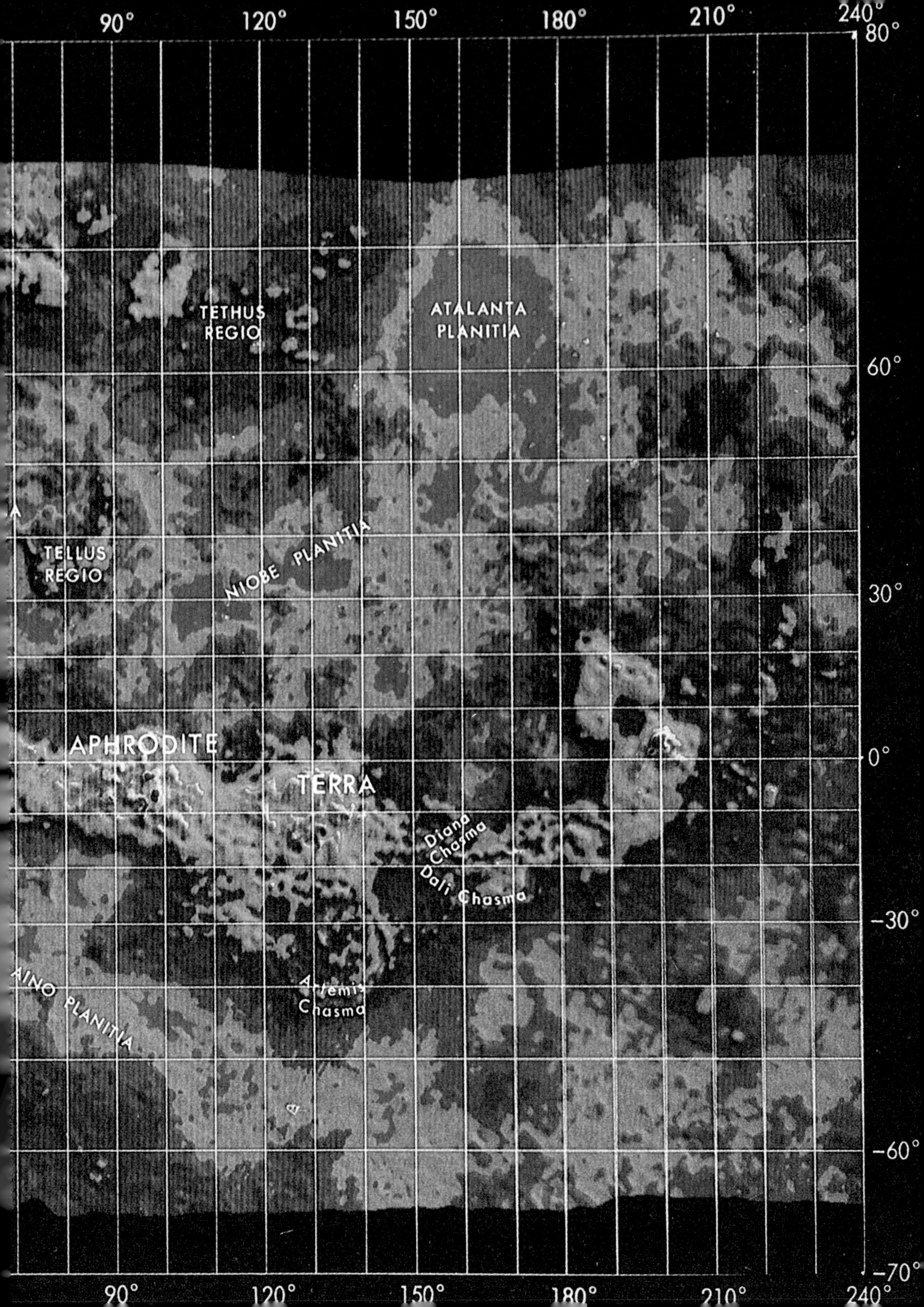

90°
120°
150°
180°
210°
240°
80°
TETHUS REGIO
ATALANTA PLANITIA
60°
TELLUS REGIO
NIOBE PLANITIA
30°
APHRODITE
TERRA
0°
Diana Chasma
Dali Chasma
−30°
AINO PLANITIA
Artemis Chasma
−60°
−70°
90°
120°
150°
180°
210°
240°

Aphrodite, as it would appear in a global view if Venus could be stripped of its perpetual cloud-cover. The different colours indicate different heights with reference to the mean surface level – the superficial impression of water is misleading, since there can be no liquid water on Venus.

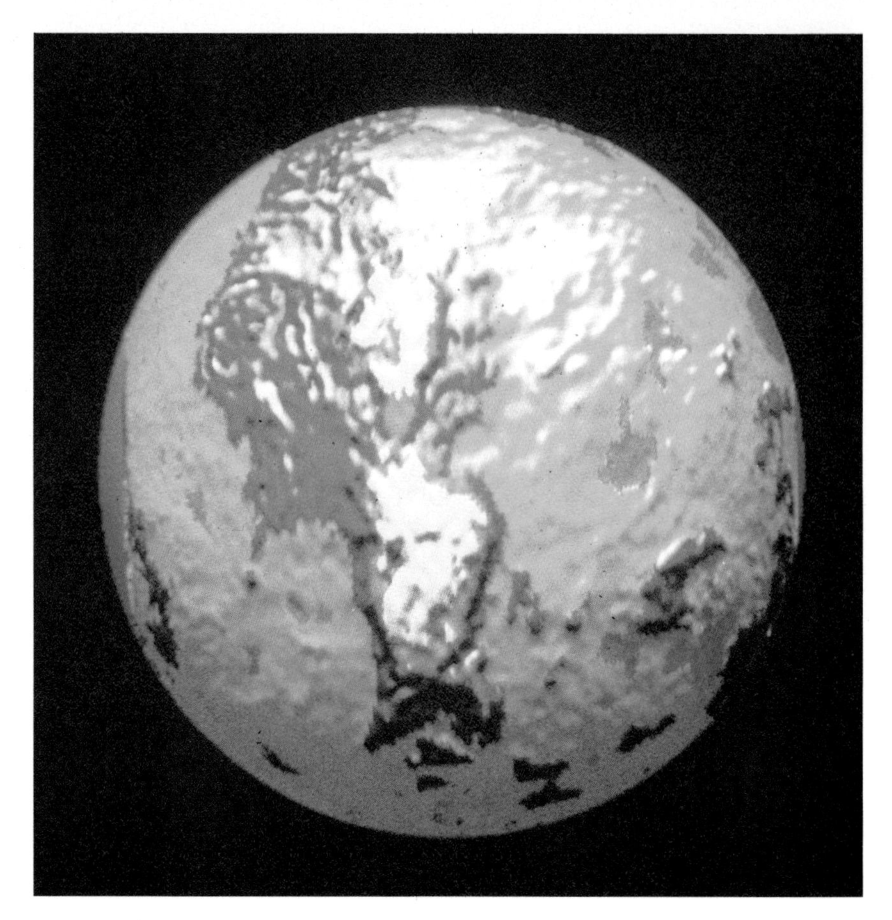

Scorpion's Tail. Lightning is almost continuous near the surface, with incessant rolls of thunder; it is particularly concentrated over Beta Regio and the Scorpion's Tail, which indicates that volcanic activity there is at least fairly frequent.

Venus' cloud region is also a shell of high-velocity winds that stretches right round the planet, above which the atmosphere is almost dead calm. It has been said that relative to ours, the cloud system of Venus is 'upside down'. On top there is a smog layer 9 miles deep; below this, extending between 35 and 29 miles above the ground, come the main clouds. It is here that we find the sulphuric acid. Sulphur compounds condense into tiny crystals, which in turn form sulphuric acid droplets and then begin to fall. Updraughts carry them back and forth, but eventually they become large enough to drop out of the cloud-layer, only to be evaporated as they enter hotter and hotter regions. The sulphuric acid rain ends at an altitude of about 20 miles, and below this comes a clear region, though of course it is still highly corrosive.

Why has Venus evolved in this remarkable way? It is so unlike the Earth chiefly because it is closer to the Sun. It used to be thought that the surface had never been cool, because carbon dioxide acts in the manner of a greenhouse, shutting in the Sun's heat, so that all the moisture was evaporated and the carbonates were driven out of the surface rocks. But the latest researches seem to paint a different picture. Analyses of the atmosphere indicate that Venus once had an ocean, at a time when the young Sun was about 30 per cent less luminous than it is now. In those days the environment may well have been Earth-like; there was plenty of water, and there seems no reason why life should not have gained a foothold – only to be wiped remorselessly out as the Sun became hotter. There was a kind of 'runaway greenhouse' effect, so that all life was destroyed, the oceans disappeared, and Venus became the furnace-like world that it is today. It is rather sobering to reflect that if the Earth had been a mere 20,000,000 miles closer to the Sun, evolution might have taken a similar course, and you and I would not be here.

If we journey in imagination through the atmosphere of Venus, the scene will be totally alien. The upper clouds have a rotation period of only four days, so that its atmospheric structure is quite unlike ours; as we descend, the light fades and the temperature rises, until at last we break through the lowest cloud-layer and see the surface below glowing redly. We hear the thunder, and see the lightning; underneath us, nothing but rocks and rubble. The Sun is never visible, because it is blotted out by the clouds, and the overall impression is remarkably like the conventional picture of hell.

Less than two decades ago we knew practically nothing about Venus; now we know a great deal – and everything we have learned tells us that Venus is a world forever beyond the reach of man.

# 5 Encounter with a Hot Planet

I have mentioned Mariner 10, which by-passed Venus in February 1974 and sent back good pictures of the cloud-tops. Yet on this mission, Venus was only incidental. The main target was the innermost planet, Mercury, and this brings me on to yet another development: the two-planet probe.

Mercury itself is a somewhat elusive world. Unless you make a deliberate search for it you are unlikely to notice it, and I would estimate that fewer than ten per cent of people in Britain or the United States have ever seen it. Yet it is obvious enough when you know where to look, and the ancient star-gazers knew it well. The main problem is that as well as being small (only slightly over 3000 miles in diameter, more similar to the Moon than to the Earth in size), Mercury is on average a mere 36,000,000 miles from the Sun, so that it and the Sun always seem to keep together in the same part of the sky. When Mercury is high above the horizon, then so is the Sun. With the naked eye, Mercury can only be glimpsed when at its best, either low in the west after dusk or low in the east before dawn.

Few telescopes will show much on its surface, and in the pre-Mariner period our knowledge of it was slight. Various observers had done their best to map the few features that could be made out, but with little success, and generally nothing can be seen apart from the characteristic phases.

Mercury: part of the surface photographed by Mariner 10 during its final active pass of the planet on 16 March 1975. The photograph was taken from a distance of 40,000 miles, 1 hour 45 minutes before closest approach; craters are shown with diameters of from 18 to 30 miles. The narrow vertical picture was 'edited' on the spacecraft and transmitted in a compressed form compatible with the receiving equipment at the Canberra station in Australia. North is at the top. This was Mariner 10's last active encounter with Mercury; before the next rendezvous the spacecraft's power had failed, and all trace of it had been lost.

In some ways Mercury was less mysterious than Venus. Its small size, low mass and feeble gravity showed that there could be little, if any, atmosphere; the surface was expected to be rather like that of the Moon, and it had no satellite. The Mercurian 'year' is equal to 88 Earth-days. The temperature was bound to be high, and the chances of finding life appeared to be nil.

Mercury never comes much within 50,000,000 miles of us, and is a difficult target. Therefore, the space planners at Houston decided to make use of Venus – and, as sheer bonus, obtain some pictures of Venus en route.

Mariner 10 was the outcome. It was launched from Cape Canaveral on 3 November 1973, and at once came under the control of the Deep Space Network at Pasadena. The usual manœuvres were carried out; Mariner was slowed down relative to the Earth, and then began its gradual swing inwards towards the orbit of Venus. It made its closest approach (about 3600 miles) on 5 February 1974, and but for this encounter it would have

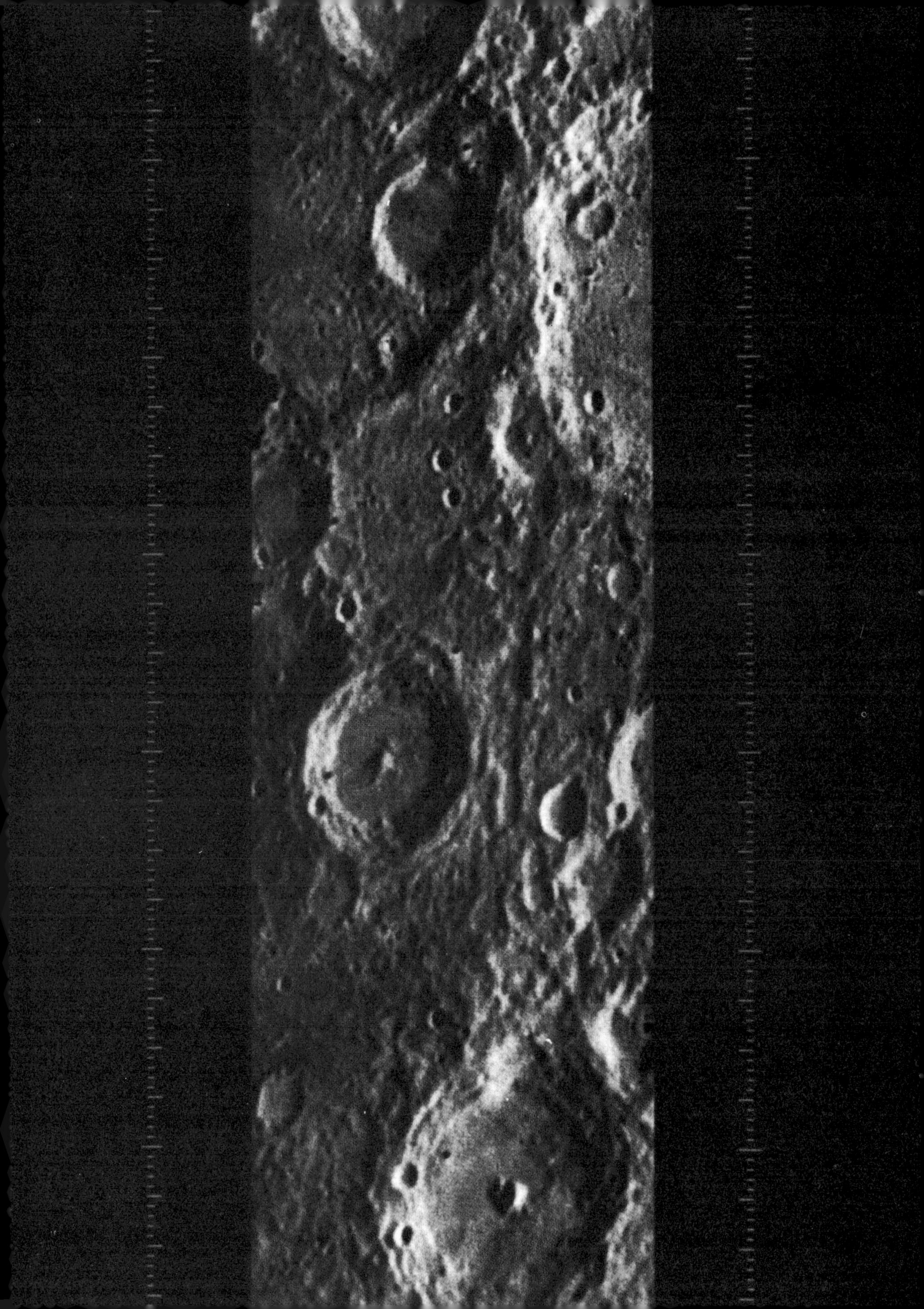

Mariner 10 trajectory, with the positions of the probe, Mercury, Venus and Earth at the launch and encounter dates. The two further active passes of Mercury were made on 21 September 1974 and 16 March 1975. The Mariner is presumably still moving round the Sun and making periodical close approaches to Mercury; it has become to all intents and purposes a permanent member of the Solar System.

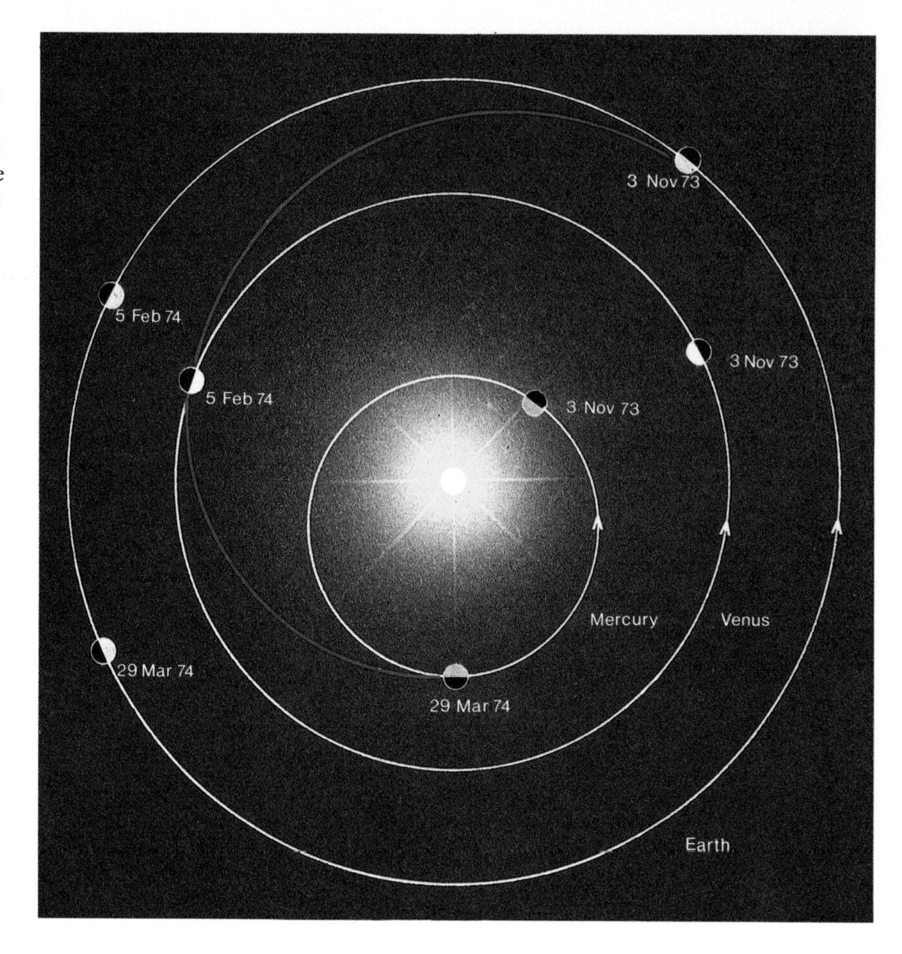


continued in an orbit that would have swung it outwards once more. However, it approached from the 'outside' and then swung back in front of Venus, so that the gravitational pull of Venus braked it still more and put it into a path bound for Mercury. This was a complicated manœuvre; nothing of the sort had been attempted before, and it is hardly necessary to add that the calculations had to be remarkably accurate.

All went well. Mariner moved inwards, and on 29 March 1974 it made its first pass of Mercury, sending back excellent pictures as well as assorted data. Over 600 photographs were received, covering wide areas of the Mercurian surface – a pleasing number in view of the fact that high-resolution pictures could be taken over a period of no more than 17 hours, because the speed of Mariner relative to Mercury was still quite considerable. Fortunately this was not the end of the mission. Mariner 10 continued in its new orbit, and encountered Mercury again on 21 September, obtaining more pictures. The third encounter took place on 16 March 1975, but by that time the equipment was starting to fail, and contact was finally lost on 24 March. Mariner is still moving round the Sun, but it is dead, and we have no hope of finding it again.

Practically all our knowledge of Mercury's surface is due to this one probe, and at the time of writing, no more missions have been announced. The only drawback was that in each of the three active encounters the same areas were presented to Mariner's cameras, so that a wide area of Mercury still remains unmapped. However, it is improbable that the unknown areas are very different from those shown by Mariner, and future missions are unlikely to provide any major surprises – though, admittedly, one can never be sure.

The first feature to be identified as Mariner 10 drew near Mercury was a bright crater, from which extended a system of rays very like those of Tycho on the Moon. A name was promptly chosen for it: Kuiper, in honour of the late Gerard Kuiper, who had founded the Lunar and Planetary Laboratory in Arizona and had been one of the great pioneers in planetary probe research.

Crater Kuiper breaks into a larger formation, now named Murasaki, and again we have the familiar lunar-type arrangement: when one crater deforms another, it is almost always the smaller crater that is the intruder. And as Mariner closed in, it became clear that the Mercurian surface really does look very like that of the Moon. There are thousands of craters, with hills, ridges, and valleys; ray systems are scattered here and there, and the only important difference is the lack of broad dark 'seas' such as the Moon's Mare Imbrium.

One particularly interesting feature is a vast ringed structure now called the Caloris Basin. The full diameter is 840 miles, and outside the mountain ring lies a complex area of ancient craters and basins which have been partly or completely flooded with lava. This region extends outwards for another 600 miles, and has no sharp boundary; it merely becomes less evident until we are back to the usual type of surface, with relatively well-marked craters and intervening smoother areas.

There was a good reason for the choice of name. 'Caloris' indicates heat, and the Basin is one of the hottest places on the entire planet. This

Ray crater on Mercury, photographed from Mariner 10. There are several prominent ray systems on the Mercurian surface, which emphasizes the outward similarity to the Moon. The main difference is the lack of large lunar-type 'maria' on Mercury; there are however areas termed intercrater plains, which may be analogous to the highland regions of the Moon.

Antoniadi's map of Mercury, regarded as the standard before the flight of Mariner 10. Earlier maps had been attempted – notably by Schiaparelli, during the 1880s – but they differed from Antoniadi's, and this was hardly surprising in view of the difficulties with which the Earth-based observer has to contend. Antoniadi named the features which he recorded, but his map proved to be so inaccurate that his names were not retained – though it is only fair to add that he was much more successful with Mars, and his Martian chart, constructed from observations made with the same telescope at Meudon, was extremely good. It is also worth noting that Antoniadi believed in the existence of a Mercurian atmosphere dense enough to support material in the form of dust, and so to produce local obscurations. This too is wrong; the trace of atmosphere around Mercury is far too tenuous to produce any visible effects.

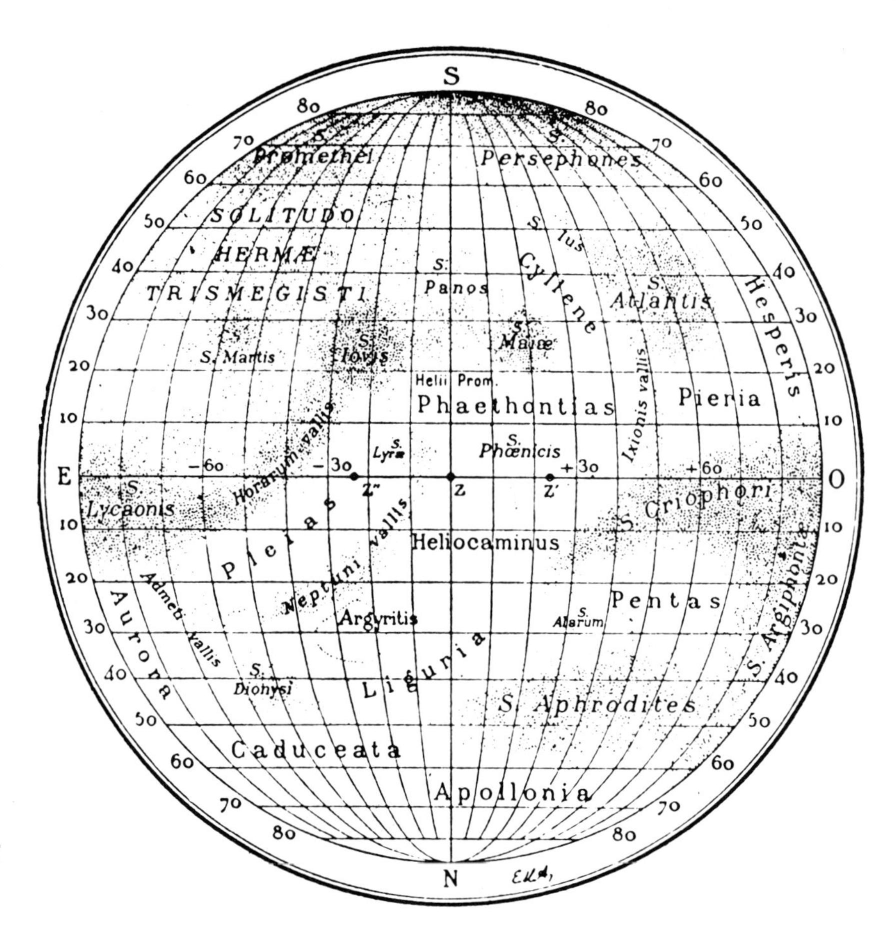


brings me on to the important question of Mercury's rotation period, because it is here that we were so very wrong before space research methods came to the rescue.

The first serious observers of Mercury were an Italian, Giovanni Schiaparelli, and a Greek, Eugenios Antoniadi. Antoniadi spent most of his life in France, and carried out his main observational work at the Meudon Observatory, outside Paris. Meudon may not be an ideal site – from it, you can see the Eiffel Tower – but it is well equipped, and the telescope best suited to planetary research is a refractor with a 33-inch object-glass. (I have a great affection for it, because during the 1950s I spent many nights using it to chart the Moon.) There are only two larger refractors in the world, both in America, and its optical quality is excellent.

Antoniadi's method was to study Mercury when the planet was high in the sky. Of course the Sun was there too, but this could not be helped, and Antoniadi constructed a map of the Mercurian surface features that he could see, or believed he could see. Before the Mariner 10 flight his map, published in 1934, was regarded as the best available.

Antoniadi, like Schiaparelli before him, believed that Mercury had a captured or synchronous rotation, with a 'day' and a 'year' of equal length: 88 Earth-days. This is how the Moon behaves with respect to the Earth, but there is one important difference. The Moon keeps the same face towards the Earth, but not towards the Sun, so that day and night conditions on the lunar surface are the same everywhere (except that from its far side the Earth can never be seen, so that nights there are darker). If Mercury had a captured rotation, it would keep the same hemisphere turned sunward. One hemisphere would be in permanent daylight, and the other plunged into everlasting night.

There would be one modification, because Mercury's orbit is not circular; it is more elliptical than ours and the distance from the Sun ranges between 29,000,000 miles at its minimum (perihelion) and as much as 43,000,000 miles at its maximum (aphelion). The orbital velocity is greatest at perihelion and least at aphelion, producing effects similar to the librations of the Moon. Between the dark and the bright regions, there would be what might be called a twilight zone, where the Sun would bob up and down over the horizon. Temperatures here would not be extreme, though on the rest of the planet they would be very unwelcoming indeed; the day side would be scorched, and the night side as cold as anywhere in the Solar System.

All this sounded plausible enough, but even before the voyage of Mariner 10 it was found to be wrong. The main evidence came from infrared measurements. Bodies give out not only visible light, but also infrared radiation, (which is simply heat), and W. E. Howard and his colleagues at Michigan in the United States found that the dark side of Mercury is much warmer than it would be if it never received any direct sunlight. Therefore, the captured-rotation theory must be incorrect.

Next came some radar investigations, this time from Arecibo in Puerto Rico. Here there is the world's largest dish-type radio telescope. It has been built in a natural basin in the ground, and obviously it cannot be steered, but it is extremely powerful, and with it Rolf Dyce and Gordon Pettengill managed to bounce radar pulses off Mercury. When such pulses are reflected from a spinning object, the echo is affected, and the rate of spin can be found. Dyce and Pettengill showed that Mercury's rotation period is not 88 days, but only 58.6.

This disposed of the twilight zone idea (to the regret of many science fiction writers), but it led to an equally curious state of affairs. The rotation period is exactly two-thirds of the orbital period. To an observer on Mercury the interval between one sunrise and the next will be 176 Earth-days, or two Mercurian years, and each time Mercury is best placed for observation from Earth the same hemisphere is turned towards us – which is what had misled Antoniadi and Schiaparelli.

The 58.6-day period was fully confirmed by Mariner 10, and we are now in a position to work out Mercury's weird calendar. At perihelion the orbital velocity is 36 miles per second, though it drops to only 24 miles per second at aphelion. Near perihelion, the orbital angular velocity is greater than the constant angular velocity of the spin, so that an observer on the surface would see the Sun move slowly 'backwards' for eight Earth-days around the time of each perihelion passage. The Sun would then be hovering over what may be termed a hot pole. There are two hot poles, one or the other of which will always receive the full blast of solar radiation when Mercury is at its closest to the Sun. One of these hot poles is in the Caloris Basin, while the other, on the

The highlands in the northern part of Mercury, photographed from Mariner 10. Craters of all kinds are shown, and the superficial aspect is remarkably 'lunar'.

opposite side of the planet, has not been studied because it was out of range during the Mariner 10 encounters.

The maximum temperature in Caloris reaches well over 700 °F. This is not as high as on Venus, because Mercury has virtually no atmosphere and so nothing to blanket in the Sun's heat, but it is certainly torrid, and if you put a tin kettle out on to the rocks of Caloris it would promptly melt.

Armed with this information, let us go on an imaginary trip to Mercury, and station ourselves first in the Caloris Basin. The Sun will rise when Mercury is near aphelion, and the solar disk will be at its smallest. As the Sun climbs towards the zenith or overhead point it will swell to its maximum diameter; it will pass the zenith and then stop, moving backwards for eight Earth-days before resuming its original direction of motion. As it drops towards the horizon it will shrink, finally setting 88 Earth-days after having risen.

An observer 90 degrees in longitude away from Caloris will have a different view. The Sun will be at its largest at the time of rising, which is also Mercury's perihelion, and after coming into view it will sink again before rising once more. It will then climb in the sky, shrinking as it does so. There will be no hovering at the zenith, but sunset will be protracted, because the Sun will disappear and then rise again briefly as though bidding adieu before finally departing, not to reappear for another 88 days.

Such are the conditions on the surface of this strange, quick-moving little world. Whether human observers will ever experience them seems doubtful. All we can claim is that Mercury is not quite so intolerable as Venus, but this is not saying much, and manned flights there are out of the question for the moment at least. Yet it is well within the bounds of possibility that an automatic station will be set up there; Mercury would be an ideal site for a solar observatory, and we would also like to know more about the material spread thinly over the whole region. Two German-built spacecraft, the Helios probes, have been put into orbits that take them fairly close in, and are being monitored by the D.S.N. at Pasadena, but a site on Mercury itself would have marked advantages.

Unfortunately the funds allotted to NASA, the United States National Aeronautics and Space Administration, have been savagely slashed recently, and no more probes to Mercury have been planned as yet. Until they are, we cannot hope to learn much more, and neither can we chart that part of the surface which was in darkness when Mariner 10 made its three active passes. But we have learned a great deal already and we now know more or less what Mercury is like, even though it can hardly be regarded as an attractive world.

# 6 Messengers to Mars

Travel to Flagstaff, in Arizona, and you will find one of the world's most famous observatories, named in honour of Percival Lowell, who founded it in 1896. Unlike so many others, it is not perched on top of a high mountain. True, it lies well above the town, but the wooded road up Mars Hill is only about a mile long, so that normally one can drive up in a few minutes. Things can be more of a problem during winter, because Flagstaff is a snowy place and more than once I have found my car slithering helplessly across pack ice, but at least there is no danger of being marooned.

Mars Hill is so named because the Red Planet was Lowell's special interest. He set up the observatory mainly to study Mars in all its aspects, and he had his own theories. To Lowell, Mars was the dwelling-place of a highly intelligent civilization, and the system of narrow, artificial-looking lines crossing the surface were genuine canals, built by the inhabitants to make up a gigantic irrigation system. Mars is desperately short of water; therefore, reasoned Lowell, every scrap must be drawn from the ice and snow at the poles.

Flagstaff was chosen because the seeing conditions there were expected to be good. On the whole this is true, even if they cannot match those of Palomar. But as well as a good site, a good telescope is needed, and Lowell accordingly installed a refractor with a 24-inch object-glass, at the time one of the largest in the world and still one of the best. Optically it is excellent, as I have good reason to know, because I carried out a great deal of Moon-mapping with it in the pre-Orbiter days. The telescope is not quite as large as that at Meudon, but it gives equally good results.

The dome of the 24-inch is not sleek and rounded, but is shaped more like a wedding-cake. The telescope itself is impressive – in its way perhaps even more so than the 72-inch and 42-inch reflectors of the Observatory, which have skeleton tubes. The 24-inch refractor seems to reach almost to the top of the dome, but there is nothing clumsy about it, and it has proved well able to adapt itself to the modern age.

The Lowell 24-inch refractor; a photograph which I took in December 1981. It may look somewhat old-fashioned as compared with the modern computer-controlled skeleton reflectors, but for planetary work it is superb.

Of course, it is not used only for planetary work. Between 1912 and 1920 one of Flagstaff's most famous astronomers, Vesto Slipher, turned his attention to the outer galaxies, and gave the first proof that they are racing away from us, so that the whole universe is expanding. But in the popular mind the Observatory will always be associated with the Solar System, and particularly with Mars.

I never met Lowell, who died in 1916, but he certainly had a strong

personality, and he has left his mark in many ways. Yet history tends to be unkind to him. He had many triumphs, but he made one great mistake, and it is for this that he is so often remembered: his firm belief that Mars must be inhabited.

A century ago such an idea was not really so outrageous. After all, Mars is much more Earthlike than any other planet. It is fairly small, with a diameter of 4240 miles, and it is further away from the Sun than we are; 141,500,000 miles on average, with a revolution period of 687 days, and an axial rotation or 'day' of 24 hours 37 minutes. There is an atmosphere, albeit a thin one, and there is water, locked up in the form of ice. The predominantly red hue of the disk is interrupted by darker patches, which are to all intents and purposes permanent, and which Lowell believed to be old sea-beds filled with vegetation. There is no dense, suffocating cloud blanket, as on Venus, and at noon on the Martian equator during summer the temperature may rise to over 50 °F., though admittedly the nights are bitterly cold. Finally, there are two very small satellites, Phobos and Deimos, neither of which is as much as 30 miles in diameter.

Lowell and his assistants used the 24-inch refractor to make thousands of drawings of Mars. They showed not only the dark features, the red 'deserts' and the white polar caps, but also numerous 'canals', stretching across the disk in a truly geometrical pattern. Lowell assumed them to be narrow rivers or piped channels, surrounded on either side by cultivated land. The important thing to note is that if Lowell's drawings had been accurate, then Mars would have been inhabited. The canal network as he drew it could not possibly have been natural.

Therefore, everything hinged on whether the canals really existed or not. Other observers, using even larger telescopes, either could not see them at all, or else showed them as broad, discontinuous streaks, not in the least like channels. I do not claim to be a good observer, but it may be of interest to show two drawings of Mars, both made with the 24-inch refractor, one by Lowell and one by myself. There is little similarity between the two.

Lowell's theories were hotly criticized even in his lifetime, because Mars did not seem to be sufficiently welcoming to support advanced life-forms, and I think it is fair to say that by 1930 or thereabouts the idea of intelligent Martians had been abandoned. Yet it was still thought likely that the dark areas were covered with vegetation, in which case Mars was not a dead world.

Before the space-probe era, we were modestly confident that we knew a great deal about Mars. The best available information indicated that there was an atmosphere made up chiefly of nitrogen, with a ground pressure of between 80 and 90 millibars, equal to the density of the Earth's air at a height of rather over 50,000 feet, and that the polar caps were made of a very thin layer of frost, only an inch or two deep at most, covering wide areas during the Martian winter and shrinking almost to nothing in the summer. The dark basins were believed to be coated with vegetation, and although there could not be any advanced plants, such as trees or flowers, there was probably plenty of more primitive organic material, perhaps comparable with our lichens or

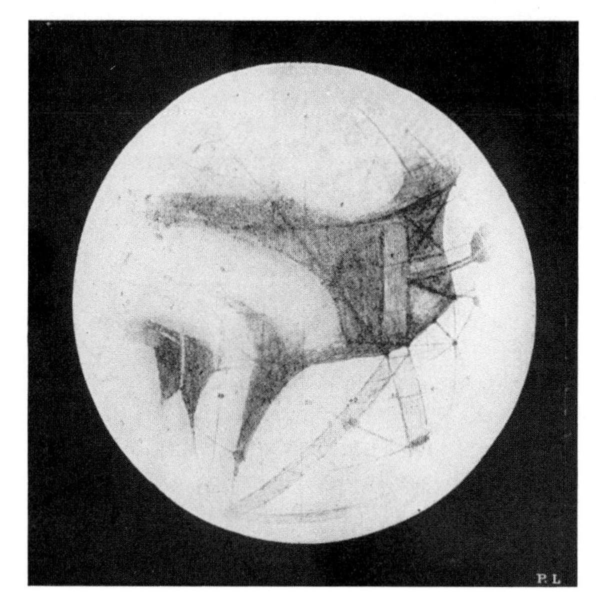


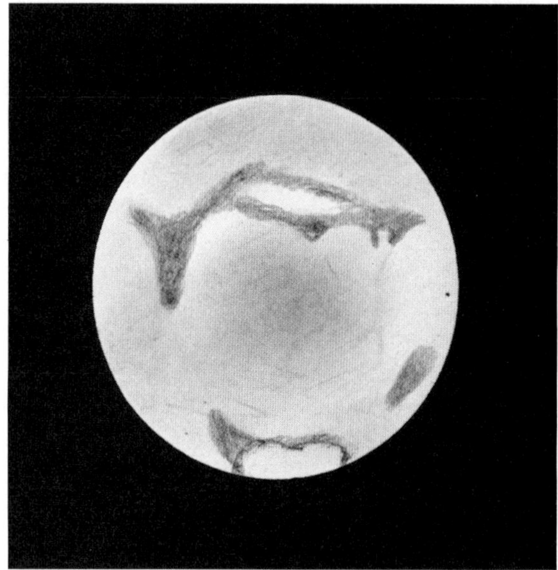

*Left:* Mars as drawn by Lowell in 1901 with the 24-inch refractor at Flagstaff. *Right:* My drawing, made on 23 February 1980, with the same telescope. Quite apart from my failure to see anything in the nature of canals, artificial or otherwise, the whole appearance is different, even though the Syrtis Major is clearly shown on both drawings – and surface changes on Mars are very minor.

mosses. Mars, in fact, was a world where life had managed to obtain a foothold, and had developed as far as it could under such adverse conditions. As for the canals – well, there was no longer any suggestion that they could be artificial; they seemed to have a basis of reality, and could be shallow valleys or even chains of low hills. Finally, it was thought that Mars was a world lacking in mountains, so that the surface was no more than gently undulating.

Within a few years most of these conclusions were found to be wrong. Our cherished theories were overturned, and it became clear that Mars was by no means the kind of world that everyone had expected. The world's largest telescopes, such as the Palomar reflector, are not generally used for looking at planets, and could in any case have added little to what we already knew. Photography is of limited use, and in fact no photograph of Mars taken from Earth can show as much detail as I can see with the modest 15-inch reflector in my own observatory. Space probes alone could give us really reliable information.

Once again the Russians made the first attempt. Mars-1 was launched in 1962, but contact with it was lost after a few weeks, so that we have no idea what happened to it. Then, in November 1964, came the American spacecraft Mariners 3 and 4. Mariner 3 was a disaster, allegedly because a technician forgot to feed a minus sign into a computer; the guidance failed, and the probe sped off into space, silent and uncontrollable. But Mariner 4 made amends in no uncertain manner.

The situation is different from that for a Venus probe. To reach Mars, the spacecraft must be speeded-up in relation to the Earth, so that it swings outwards, meeting Mars at a pre-computed position. Mariner 4 lifted off from Canaveral on 28 November; the two steps of the launcher broke away on schedule, and Mariner, now on its own, was put into a transfer orbit, bound for Mars. Within two days it was half a million miles from Earth, and it coasted along, with an occasional course correction, until the Mars encounter on 14 July 1965. The distance between the probe and the planet was then only 6118 miles, but both were 134 million miles from Earth.

The journey had taken almost eight months, and had involved a total distance of 418 million miles, which seems a long way when we remember that at its closest Mars can approach us to within 35,000,000 miles; but there was no other possible orbit, because of the fuel limitations. After the fly-by, Mariner 4 continued on its way, and was tracked for some time later even though its main work was done.

Since the mission was a first attempt, excluding the abortive flight of Mariner 3, its success was truly amazing. There were so many factors to be taken into account. For instance, the spacecraft had to be correctly aimed towards Earth, otherwise no messages from it could have been received and this was done by using two reference objects, the Sun and Canopus. The best way to explain it is to picture a weight that is hung from a cord. It will tend to spin, but a second cord, fixed to it at about a right angle, will steady it. With Mariner, the Sun was used as one 'lock', and the second was Canopus, which is bright enough and isolated enough in the sky to be particularly suitable. Mariner's sensors located Canopus, admittedly after a day of searching around, and the method worked well, though on one occasion the sensors lost Canopus and locked temporarily on to the wrong star. Power for corrective manœuvres was drawn from the solar paddles that gave Mariner its distinctive appearance.

It is by no means easy to transmit information back across so great a distance. Each picture of Mars took 8 hours 35 minutes to build up on the screens at the D.S.N., and since 21 photographs were sent back in all the process was a slow one. Moreover, the total power received from Mariner was only 0.0000000000000000001 of a watt (if you care to count the number of zeros to the right of the decimal point, you will find that there are 18 of them). The fact that all this became possible a mere eight years after the first American artificial satellite, Explorer 1, shows how rapidly things had progressed.

I was in Pasadena when the Mariner 4 pictures came through, and like everyone else I was taken aback. The Martian surface was not smooth. There were obvious craters, and the dark tracts did not seem to differ from the red regions except in colour, so that the idea of sunken basins filled with vegetation was incorrect. In some ways, Mars resembled the Moon more than the Earth, though it was not really much like either. The best picture, the eleventh of the series, showed a large crater in the tract that had been named Atlantis, 30° S. of the Martian equator. One of the peaks on its wall was estimated to rise 13,000 feet above the floor, and nearby smaller craters were scattered around.

Spectacular though the pictures were, they did not represent the most important part of Mariner's programme. This, undoubtedly, concerned the Martian atmosphere, which proved to be unexpectedly thin. The method used was highly ingenious. Shortly after 02.19 hours G.M.T. on 15 July, more than two hours after closest approach, Mariner went directly behind Mars. Of course it could not be seen – it was hopelessly out of telescopic range, and had been so almost from the very beginning of its flight from Canaveral – but its radio signals were strong, and just before the occultation behind Mars these signals came to us after having

Mariner 4 launch vehicle at Complex 12 at the Kennedy Space Centre, Cape Canaveral. This probe was the first to orbit Mars and provide good close-range pictures. Presumably it is still moving round the planet, though its power failed before the end of 1965.

passed through the planet's atmosphere. The way in which the signals were distorted gave a vital clue as to the atmospheric density, and there was another chance when Mariner emerged from behind Mars at 03.13 hours. The occultation itself had lasted an hour; the distortion of the signals due to the Martian atmosphere was traced for two minutes before occultation and another two minutes afterwards.

In some ways the results were rather disappointing. Instead of having a ground pressure of over 80 millibars (as against about 1000 millibars for the Earth), the pressure turned out to be less than 10 millibars, so that it corresponded to what we normally call a laboratory vacuum, and was far too rarefied to support much in the way of life. Moreover, there was absolutely no sign of any canals, artificial-looking or otherwise.

Things were not improved by the next two American probes, Mariners 6 and 7, which were dispatched in the spring of 1969 and encountered Mars on 31 July and 4 August respectively, less than a fortnight after Neil Armstrong had stepped out on to the surface of the Moon. The pictures received were much better than those from Mariner 4, and there were more of them; the areas covered were different, but the overall impressions were similar. Craters were plentiful in some regions, but less so in others, where the landscape was described as 'chaotic'. The thinness of the atmosphere was confirmed, and the main constituent was found to be carbon dioxide rather than nitrogen. The final *coup de grâce* was given not only to the canals, but also to the idea that the dark regions were ancient sea-beds; in fact the most prominent dark marking on Mars, the V-shaped Syrtis Major, turned out to be a lofty plateau, sloping off to either side. Another feature, Hellas, had always been regarded as elevated, and perhaps snow covered at times; instead, it proved to be a basin.

It seemed as though our opinions about Mars were being turned upside-down, but the overall aspect was of a barren, cratered waste, just as desolate as the Moon and probably less interesting. Even the nature of the polar caps was in doubt. Rather than being made up of frost, they could be solid carbon dioxide, the same as the so-called dry ice found in an ice-cream seller's barrow. This was an old theory that had fallen into disfavour; after Mariners 4, 6 and 7, it was revived.

Craters were evident even in the polar zones, notably one shown on a Mariner 7 picture which was nicknamed the Giant's Footprint. At this time the temperature at the pole was as low as –190 °F.

The flights of 1969 ended what may be termed the second phase in the rocket exploration of Mars. At that time nobody could tell that the early results had been misleading, and that by sheer bad luck all the successful Mariners had passed over the least interesting areas anywhere on the planet.

The year 1971 was important in Martian history, and four probes were launched towards the planet – two American and two Russian, though only one of them was successful. Their aims were not the same. The American vehicles, Mariners 8 and 9, were designed to enter closed orbits around Mars, so that they could continue sending back photographs and assorted data for several months instead of only a few days. The Russians were more ambitious. Each probe was made up of an

Mars, photographed by the orbiting section of Viking 2 in August 1976 as it approached the dawn side of the planet. At the top (north), with water ice-cloud plumes on its western flank, is one of the great volcanoes of Tharsis: Ascræus Mons. The great rift valley Valles Marineris is seen in the middle of the picture, with the large, frosted basin Argyre I near the bottom. At this time Viking was 261,000 miles from the planet.

orbiter, which would circle Mars, and a lander, which would break free and make a controlled landing on the surface.

Mariner 8, first of the four vehicles, was a prompt failure. The second stage of the launcher failed to ignite, and Mariner made an undignified descent into the sea.

Next it was the turn of the Russians, but again they achieved very little. Both probes managed to enter orbits round Mars, and each ejected a lander, but the first one crashed out of control, while the second had barely started to send back data when it stopped transmitting permanently. The cause of this sudden silence has never been explained. Either there was a straightforward power failure, or else the grounded vehicle was toppled over by a dust-storm or a violent gust of wind. (My suggestion that a Martian conservationist had crept up and switched it off was not taken very seriously!) The probe is still presumably lying where it fell, in the region between the two bright areas of Electris and Phæthontis, so that some day it may be recovered and examined.

Mariner 9 caused yet another complete change in our outlook. The launch date was 30 May 1971, and by mid-November it had reached the neighbourhood of Mars, incidentally taking the first good pictures of the two miniature satellites, Phobos and Deimos. After some well-controlled manœuvres, Mariner 9 entered a path which took it round Mars once every 11 hours 58 minutes, at a minimum distance of 850 miles

Ice on Mars. This photograph was taken by the Lander of Viking 2 from Utopia on 18 May 1979 and was relayed to Earth via Viking Orbiter 1 on 7 June. It shows a thin coating of water ice on the rocks. The build-up of frost was exactly the same as that which had occurred one Martian year previously, when the ice remained on the surface for about 100 days. It is believed that dust particles in the Martian atmosphere pick up pieces of solid water. This combination is not heavy enough to settle on the ground; but carbon dioxide, which makes up most of the planet's atmosphere, freezes and adheres to the particles, so that they become heavy enough to sink. After they have reached the ground, the warmth of the Sun during the day evaporates the solid carbon dioxide and returns it to the atmosphere, leaving behind the water ice and dust. The layer of ice shown here is probably no more than 1/1000 of an inch thick.

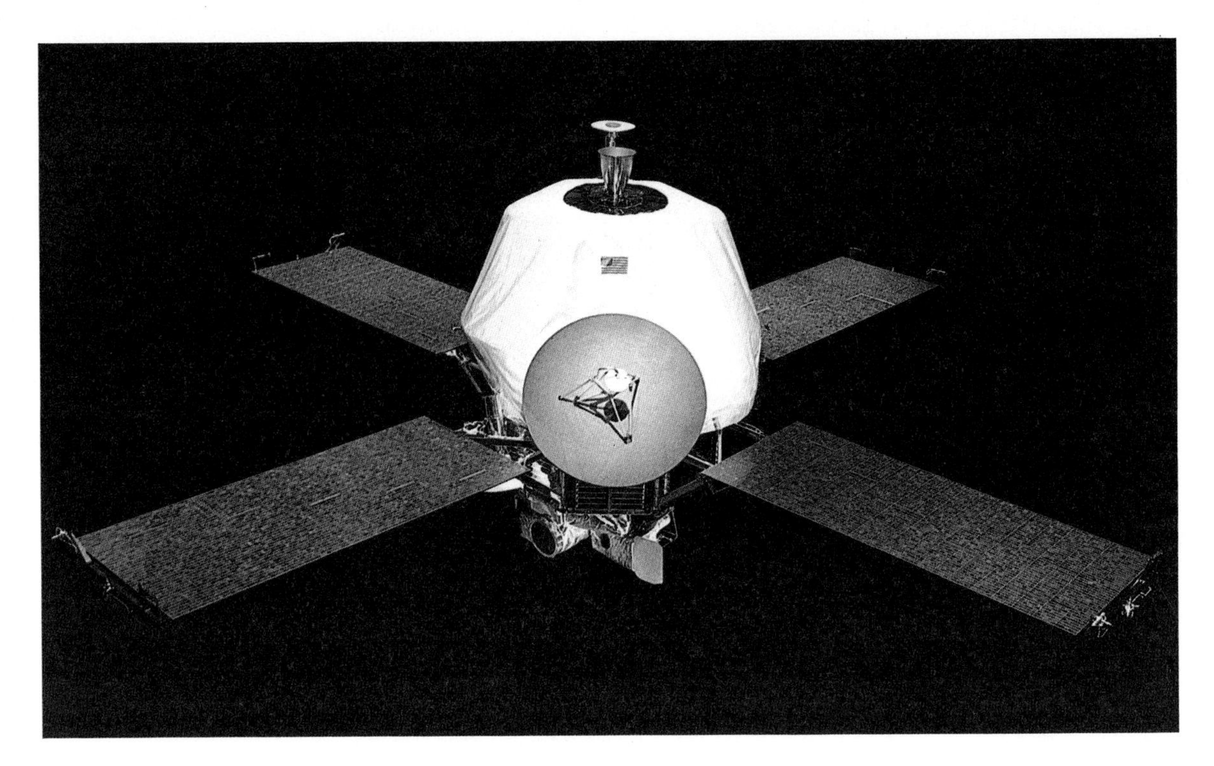

Mariner 9. The 2200-pound spacecraft is shown with the green, circular, high-gain antenna facing the viewer. The white shroud is a thermal blanket covering the 300-pound thrust retro-engine. The rocket nozzle protrudes from the top, with the low-gain antenna behind it. The scan platform carrying the scientific experiments is visible below the spacecraft. Mariner measures $7\frac{1}{2}$ feet from the scan platform to the tip of the low-gain antenna, and spans 22 ft $7\frac{1}{2}$ inches from panel tip to panel tip.

above the ground. So far, so good; but when the first pictures were received, they showed little apart from four rather large, obvious spots.

The reason for this was by no means mysterious, and it was something that I had expected from my own observations. Using my 15-inch reflector, I had been making a series of drawings of Mars under fairly good conditions. Up to the end of September the usual features were seen with perfect clarity, but then the details started to fade out, and by the end of the first week in October the disk was more or less blank. Mars was experiencing one of its global dust-storms, when the atmosphere becomes opaque and the surface features are completely masked. It was well into the New Year before the dust cleared away, but as soon as conditions reverted to normal Mariner 9 began to transmit pictures which were as surprising as they were dramatic.

Mars was not merely a world of craters, ridges, hummocks and dunes. There were lofty volcanoes, deep valleys, and amazingly complex systems of chasms unlike anything either on the Earth or on the Moon. The volcanoes were particularly striking. Several of them lay on a high ridge in the area known as Tharsis; one of them, now called Olympus Mons (Mount Olympus) has a base over 300 miles across, while the summit, towering to 15 miles above the general surface level, is crowned by a 40-mile in diameter caldera. Olympus Mons is a shield volcano, like those on Hawaii but much larger and more massive. It is not unique; there are other volcanoes hardly less imposing, both in Tharsis and elsewhere, together with mountains of less obviously volcanic type.

The Valles Marineris or Mariner Valley, extending from the Tharsis ridge, is over 3000 miles long, three to four miles deep in places, and 125 miles broad at its widest part. Near sunrise or sunset, when one end of it is

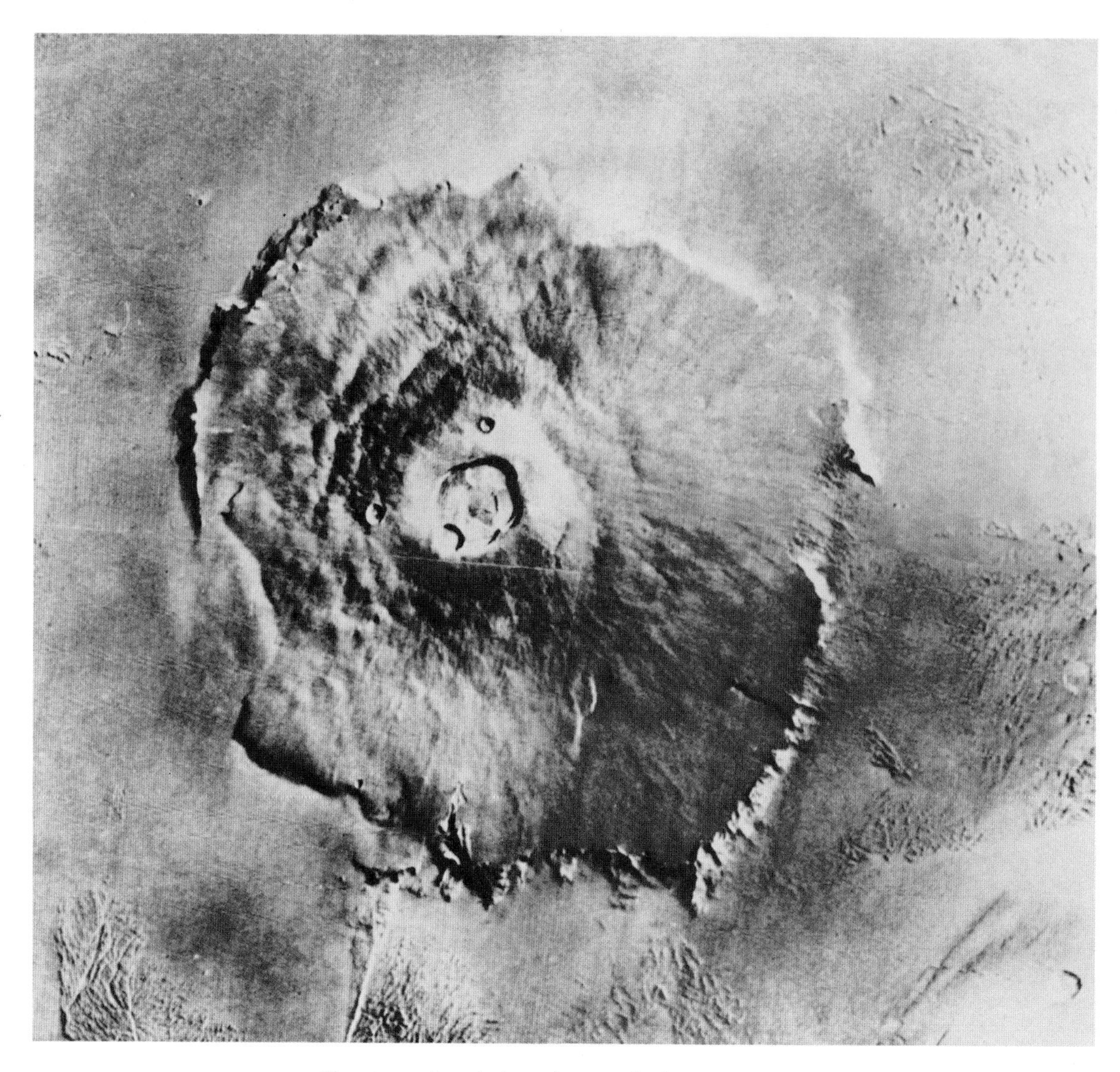

The highest volcano in the Solar System: Olympus Mons, which is the supreme example of a Martian shield volcano. This Mariner 9 picture shows the great escarpment round the base of the volcano, about 2400 miles long, which has been compared with a wave-eroded sea cliff on a terrestrial volcanic island. The escarpment is sharp for more than half its length; in a few places it is absent, probably covered by lava flows. No definite signs of activity have been found.

illuminated and the other in darkness, temperature gradients must make winds shriek through it in the thin Martian atmosphere, and it is associated with tributaries that look very much like a system of old river-beds. South of the equator there is a labyrinth of canyons generally nicknamed the Chandelier, any one of which would make our Grand Canyon of Arizona seem very puny. Many of the channels and canyons appear to be drainage systems, and the evidence in favour of past running water on Mars is overwhelming.

And yet there is a paradox here. There is no sign of marked erosion, even though the Martian atmosphere is dusty, and dust – even fine dust – is highly abrasive. The channels and the craters do not look as though they have been filled up, so that they can hardly be very ancient; tens of thousands of years perhaps, but not millions. In this case, Mars must have had liquid water on its surface not so very long ago. Water cannot exist there now, because the atmospheric pressure is too low. It follows that in its more fertile period, Mars must have had a much denser atmosphere than it has today.

The Martian escape velocity is low, only 3.1 miles per second compared with 7 miles per second for the Earth and over 6 miles per second for Venus. Any Martian atmosphere of the same type as ours would therefore have been lost very early on, around 4000 million years ago or even before, and we have to explain why there is such strong evidence for a dense atmosphere in much more recent times.

Can the polar caps provide an answer? Mariner 9 demonstrated that the main caps are made of ordinary ice, though during the Martian winter there is also a coating of solid carbon dioxide, and the north and south caps do not seem to be identical. At any rate the caps are thick, perhaps more so than those of our Arctic and Antarctic. This has led some investigators to suggest that they may provide the clue to the long-term climatic changes on the planet.

The axial inclination of Mars is much the same as ours: between 23° and 24° to the perpendicular, and, again as on Earth, summer in the southern hemisphere occurs near perihelion. On Earth the effects are not very pronounced. In theory the southern summers are shorter and hotter than those of the northern hemisphere, and the winters are longer and colder, but with us the greater amount of sea in the south acts as a stabilizer. Not so with Mars, where there is no sea, and the orbit is much more elliptical. The extremes of temperature are much greater in the southern hemisphere, and the extent of the cap is much more variable.

The axial tilt of Mars is not always the same. It ranges between 13° and as much as 35° over a cycle of about 100,000 years, and when at its most extreme, it is suggested that the southern cap may melt during the summer, releasing water vapour and thickening up the atmosphere. Mars would then enjoy a more congenial climate, though it would not last for long. If this theory is sound, we may have an explanation for the past presence of running water. Alternatively, it may be that there are occasional periods of intense vulcanism, when gases are vented from Olympus Mons and other volcanoes. Whether they are active, dormant or extinct is something which we still do not know.

For almost a year Mariner 9 continued to transmit, sending back a grand total of 7329 high-quality pictures before contact with it was finally lost on 27 October 1972. Had the Martian scene been one of dull craters and formless rubble, as had been expected, enthusiasm would have been dampened. As things were, the next developments were awaited eagerly. There seemed at least a reasonable chance of discovering Martian life, but this was one problem that Mariner 9 could not solve, and four more Russian attempts in 1973 were abortive; two of the probes went out of contact, and the other two missed Mars completely. The main hopes were concentrated on the two new American Vikings, which were scheduled to make controlled landings and carry out on-the-spot analyses of the Martian crust.

The Vikings were identical twins. Each consisted of an Orbiter and a Lander, which would travel together across space and enter a closed path round Mars in the same way as Mariner 9 had done. Then, at a given command, the Lander would be separated, and would come down

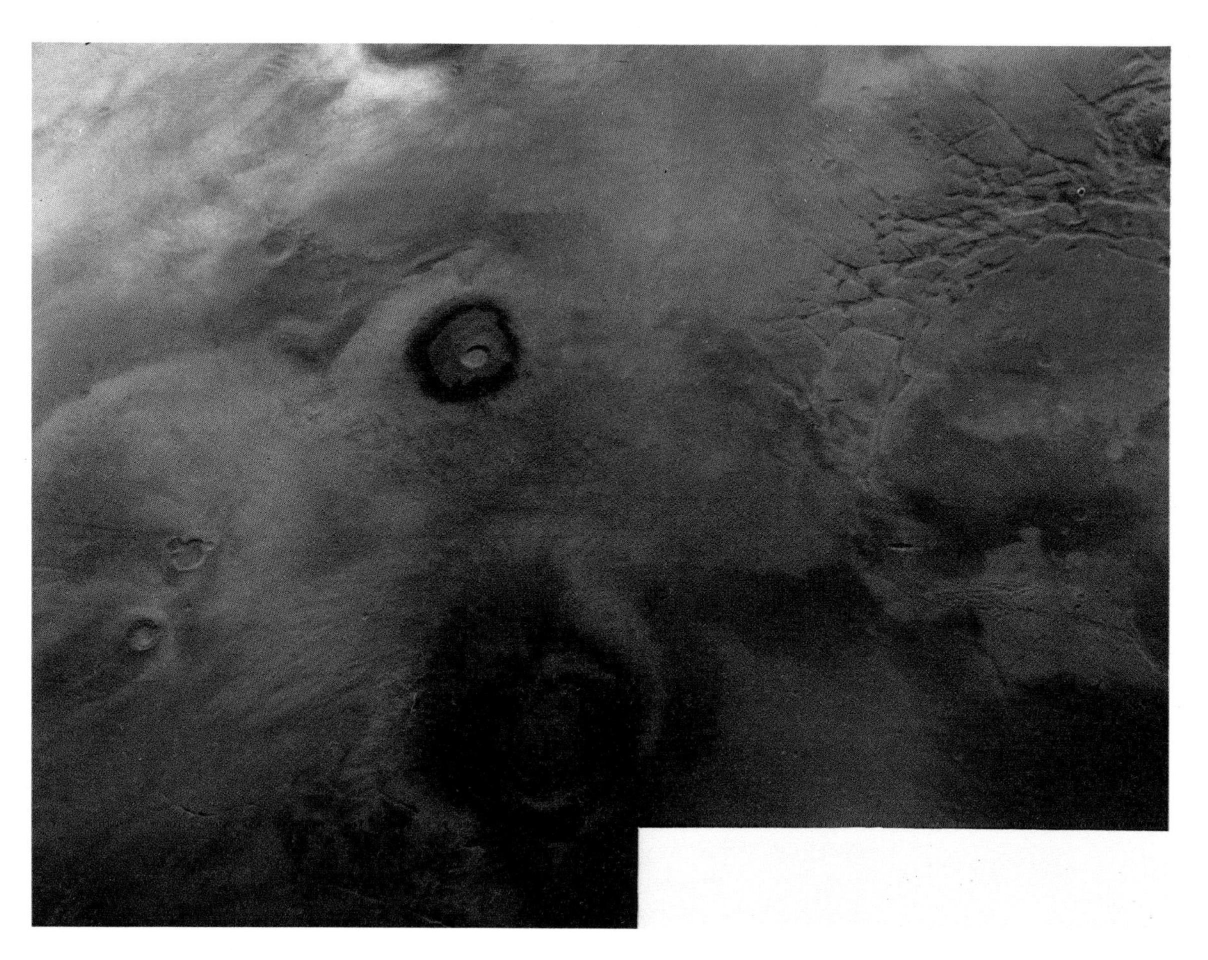

An oblique view along the Tharsis Ridge, taken on 13 July 1980 by the Viking 1 Orbiter. The three Tharsis volcanoes Arsia Mons, Pavonis Mons and Ascræus Mons (from bottom) rise to an average height of 10.8 miles above the mean surface level of Mars. The smaller volcanoes. Biblis Patera and Ulysses Patera are seen to the left; to the right, the western part of Noctis Labyrinthus (the 'Chandelier'), a region of complex faults. The upper portion of the mosaic is dominated by clouds, and the valleys to the lower right are filled with haze. The photograph covers an area of 1240 miles by 1550 miles, i.e. about 2,000,000 square miles.

through the atmosphere, braked partly by parachute and partly by rocket power. The situation was slightly easier than on the Moon, because the Martian atmosphere is dense enough to make parachutes useful even though they cannot cope with the slowing-down process on their own. On the other hand the landing sequence could not be stopped once it had been started, and there could be no second chance.

Viking 1 began to orbit Mars on 19 June 1976. Its first task was to survey the proposed landing-site in the ochre tract of Chryse, the 'Golden Plain', and to continue the various programmes started with Mariner 9. The pictures sent back were superb; besides the volcanoes, valleys and craters there were extensive lava flows, dunes and systems of canyons, as well as one or two exceptional features that caused some discussion; one mountain clump bore an uncanny resemblance to a human face! The landing-site had to be surveyed with special care, because if the Lander came down in an unsafe area, or if it struck a rock and tilted over by more than 19° from the perpendicular, it would be unable to transmit.

Alarm signals were sounded from Arecibo in Puerto Rico, where the 1000-foot radio telescope had been making radar checks of the Chryse area. There seemed to be an unacceptable degree of roughness, and the Orbiter pictures were not encouraging. Eventually the first site was rejected, and a second selected, still in Chryse but further northwest. This appeared to be better, but the final choice was still further west, and the die was cast. On 20 July the attempt began.

The capsule which separated from the Orbiter, was made up of three main sections: the Lander itself, a cone-shaped aeroshell of aluminium alloy, and a base cover. The Lander was enclosed to prevent damage by frictional heating as it plunged through the atmosphere, while the braking engines were in the aeroshell and the parachute was in the cover. When the signal from D.S.N. at Pasadena was received, the Orbiter separated faultlessly; for three hours the Lander glided down towards Mars, until by the time it reached the top of the resisting part of the atmosphere it was travelling at 10,000 m.p.h. The heat-shield did its work well, and at 19,000 feet from the surface the velocity had dropped to only 1000 m.p.h., at which point the parachute was deployed and the aeroshell fell away. At just below 4000 feet the parachute too was jettisoned, and the braking rockets came into play. Seconds later the Viking landed in Chryse at a speed of less than six miles per hour, only 20 miles from the planned position. Luck was on NASA's side; the Lander touched down only 25 feet from a boulder which would have caused fatal damage if the landing had been on top of it.

Actually, the landing had been completed before news of it could be received in Pasadena, because the signals from Mars took nearly 20 minutes to reach the Earth, but the feeling at D.S.N. was one of triumph. Viking was the most ambitious spacecraft ever launched up to that time, and from a purely technical point of view the mission was probably more difficult than that of putting a man on the Moon.

The first pictures from Chryse were received almost at once. Viking was standing in an orange, rock-strewn plain; the pads had penetrated less than two inches into the soil, so that clearly the surface was reassuringly solid. The sky was pink, not dark blue as many people had expected. The density of the Martian atmosphere is too low to scatter the sunlight and make the sky blue, as on Earth, but the fine dust causes a general pinkness during the day-time. The temperature was far below freezing-point; there were no visible clouds, and the windspeed was less than 15 knots.

Viking 2 followed, and its Lander reached Mars on 3 September, 4600 miles away from Chryse and further north, in the plain of Utopia. The overall scene was similar, and since the experiments carried out by the two probes were of the same kind it seems best to discuss them together.

The atmosphere turned out to be 95 per cent carbon dioxide, with about 2 per cent of nitrogen and smaller quantities of other gases, including argon and oxygen. The pressure is only about 7 millibars, which is equivalent to the density of the Earth's air at a height of over 120 miles above sea-level, but there are seasonal variations; during the southern winter some carbon dioxide is frozen out of the atmosphere and deposited on the polar cap, so that the atmospheric pressure is temporarily reduced by a measurable amount (the same thing happens, to a rather lesser extent, during the winter of the northern hemisphere). Windspeeds recorded by the Landers were generally light, though it has since been found that they can rise to at least 100 knots even though they have little force. The colour of the landscape is ochre, and when I saw the pictures I was reminded of the Painted Desert of Arizona, which, as Percival Lowell himself had pointed out long before, does look remarkably Martian.

Analyses of the surface rocks showed that they are volcanic, and made up of the usual substances such as silicon, iron and calcium. At the Viking sites there was rather more water vapour than had been anticipated, though not nearly as much as the Orbiters had found over the polar caps and over sunken basins such as Hellas and Argyre.

Each Lander was equipped with a scoop, together with what can best be described as a portable chemical laboratory. The scoop was extended, and grabbed samples of Martian material, drawing them back into the main spacecraft for analysis. The idea was to 'feed' the samples with nutrients, and see whether the reactions showed any trace of organic material.

The first sample from Viking 1 was collected on 28 July, and the analysis was carried out. The findings were inconclusive. Reactions were shown which might have been due to living matter, but were more probably purely chemical. Later results, from both Landers, gave the same story, and the most we can say at the moment is that no definite signs of Martian life have been found. Most of the investigators have now

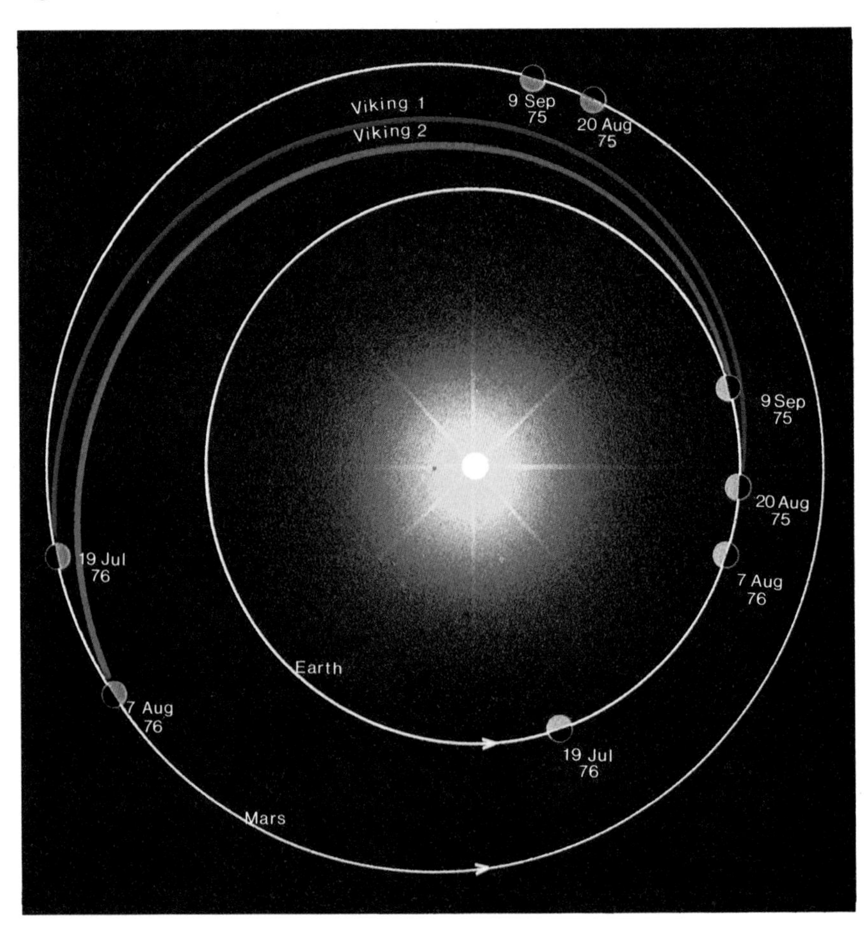


The trajectories of Vikings 1 and 2, and the relative positions of the Earth and Mars at launch and orbit insertion dates. The landing section of Viking 1 came down in Chryse on 20 July 1976; Viking 2, in Utopia on the following 3 September. Diagram by Paul Doherty

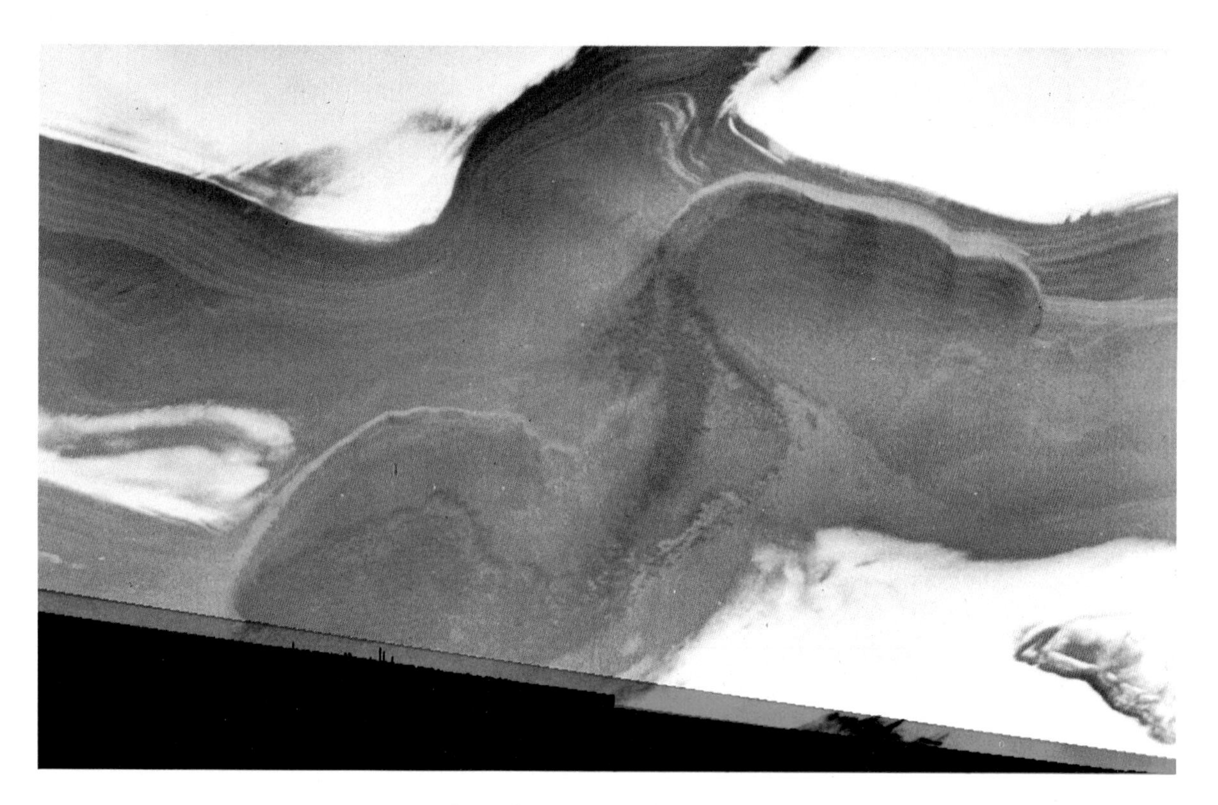

The north polar region of Mars as shown from Viking 2 Orbiter on 26 October 1976. It is midsummer; the seasonal coating of solid carbon dioxide at the pole clears, to reveal water ice and layered terrain beneath, highlighted by frost. Contact between ice and ground at the top occurs at the brink of a scarp 1650 feet high; steps on the scarp face are 165 feet thick. The regularity of the layers suggests periodical changes in the orbit, and hence the climate, of Mars. The scarp is apparently an erosion feature and the forms of the arcuate cliffs show the complex nature of erosion here. As elsewhere around the north polar cap, there are dune-like features with rippled texture. Material of the dunes may come from the eroded layered terrain. Photograph taken from 1370 miles.

come to the reluctant conclusion that life does not exist.

We cannot be certain that Mars is sterile, but the current evidence does point that way, and we can learn no more from the Vikings, though the first Orbiter is still capable of transmitting some data and may continue to do so for some years. In any case, one long-standing question has been answered: the canals do not exist. A few of them shown in the drawings by Lowell and others fit in reasonably well with valleys and chains of craters, but frankly I think that this is coincidence, and that the whole geometrical *réseau* was caused by tricks of the eye. The brilliant-brained Martians have been finally banished to the realm of science fiction.

Yet can we safely dismiss the idea of past life on Mars? There is plenty of water locked up in the form of ice, and if Mars were formerly less forbidding than it is now, it is conceivable that life appeared there. On the other hand the fertile periods must be brief, and life is notoriously slow to evolve. Martian fossils may or may not be found; I admit to being pessimistic, but it is fair to say that we may be seeing the planet at its very worst.

Phobos and Deimos, the two satellites, were discovered as long ago as 1877, but before the Mariner and Viking flights we knew nothing about their surfaces, because both are so small that they look like mere specks of light in our telescopes. Each is irregular in shape, and pitted with craters, ridges, and (in the case of Phobos) strange grooves. They are close to the planet; Phobos orbits at only 3700 miles above the surface, and completes one revolution in 7 hours 39 minutes, so that to a Martian observer it would rise in the west, cross the sky in a mere four-and-a-half hours and set in the east. Both are quite unlike our massive Moon, and may not

The 'scoop' sent out from Viking Lander 1, to collect Martian material and draw it back into the main probe for analysis. There was an initial crisis because a latchpin jammed and prevented the collecting operation from being completed. The planners decided to extend the boom beyond its first position, and this proved to be successful; the awkward pin was released, and fell to the ground, where it was subsequently photographed. (It would have been ironical if the whole operation had been wrecked by a 3-inch pin!) On 28 July, the eighth Martian day or sol, the grab secured a sample, and also dug a trench 3 inches wide, 2 inches deep and 6 inches long.

be genuine satellites. Beyond Mars we come to the asteroid belt, in which move thousands of small bodies; it seems likely that Phobos and Deimos have come from there, in which case they were captured by Mars in the remote past. Neither would be of much use as a source of light during the Martian night, and Deimos would look rather like a large, dim star.

Our present view of Mars is completely different from that of pre-1965 days. Yet there is still much that we do not know; for instance, do any of the volcanoes of Tharsis show traces of activity? and why are the dust-storms so sudden and so all-concealing? At the moment there are no spacecraft keeping watch on Mars, so that there is still work for telescopic observers to do.

This brings me back once more to the Lowell Observatory, which is the centre of the International Planetary Patrol. Observers at Flagstaff, Meudon, Mauna Kea and elsewhere are continuing to monitor Mars, both visually and photographically, so that if there are any major outbreaks of vulcanism we will be able to follow them. Many thousands of observations are stored at Flagstaff, and many more will be needed before we can claim that we really know the true nature and history of Mars. It is fascinating work – and if I may cite a personal example, I well remember the feeling of excitement I had in February 1980, when I was using the 24-inch refractor with Dr Charles Capen, and we suddenly saw a rift in the northern polar cap that had not been recorded since 1888.

Mars, alone of the inner planets, is not hopelessly hostile, and the presence of ice means that it will be able to provide future explorers with materials much more valuable than anything to be found on the Moon. It must be the next target for manned expeditions, and from a purely technical viewpoint, a journey there might become possible within the next few decades. Motivation is needed; and to quote Eugene Cernan, 'If Viking had shown a little green man with long ears looking out, we'd be on our way tomorrow!' But, if humanity refrains from wiping itself out, the journey will be undertaken sooner or later.

If colonies are established on Mars, they may become self-supporting.

Whether children born and brought up there, under conditions of only one-third of the Earth gravity, would ever be able to come to their parent world is by no means certain, and we may reach a situation in which there are two entirely different branches of mankind. Admittedly, this is looking well into the next century or even beyond, but if we are to establish our civilization on another planet it must surely be on Mars.

# 7 The Giants

Jupiter, from Voyager 1; 1 February 1979, from a range of 20,000,000 miles. Objects as small as 375 miles across can be seen on this picture. The Great Red Spot is excellently shown. Note also the change in surface appearance near Jupiter's poles, indicating that the conditions there are different from those in lower latitudes.

For most of the time, the Jet Propulsion Laboratory at Pasadena, with its Deep Space Network, is a calm place. Work is going on all the time, quite apart from the ceaseless monitoring of the probes far away in the Solar System, but there is little hustle and bustle. Yet there are times when the whole Laboratory, outside the D.S.N. itself, seems to buzz like a beehive. This happens whenever a particularly interesting spacecraft makes a close rendezvous with one of the planets. Newsmen are there in force; there is a separate room made entirely available for the Press, with television screens, desks and telephones; regular bulletins are given, and there are daily news conferences in the Von Kármán Auditorium. NASA does not believe in secrecy, and all important information is released as soon as it has been checked and verified.

Moreover, each new encounter seems to provide its quota of surprises. Before the spacecraft era, who would have believed in the sulphuric acid clouds of Venus, the fiercely hot Caloris Basin of Mercury, or the Martian volcanoes of the Tharsis Ridge? But perhaps the greatest shocks came not from the inner planets, but from the giants of the Solar System, Jupiter and Saturn. I was at the J.P.L. for several of these encounter periods, and some of the discoveries seemed to be incredible. One had to do one's best to emulate the White Queen in *Through the Looking-Glass*, who made a daily habit of believing at least six impossible things before breakfast.

The story of the missions to the further reaches of the Solar System began in 1972, with the launching (from Canaveral, of course) of two ambitious probes, Pioneers 10 and 11. Both carried numerous scientific experiments, but their main tasks were in connection with Jupiter, a gigantic world over 80,000 miles in diameter, big enough to hold more than a thousand globes the volume of the Earth and more massive than all the other planets put together. As seen from Earth, it is extremely bright, even though it is so far away. Its mean distance from the Sun is 483 million miles, and its 'year' is $11\frac{3}{4}$ times longer than ours.

Telescopically, Jupiter shows up as a yellowish, decidedly flattened globe crossed by streaks known as cloud belts, and with other less prominent features such as spots, wisps and festoons. The most casual glance is enough to show that it is totally un-Earthlike, and it has long been known that the surface is not solid, but is made up of gas, mainly hydrogen and helium together with unpleasant hydrogen compounds such as methane (marsh-gas) and ammonia. It spins very quickly – a

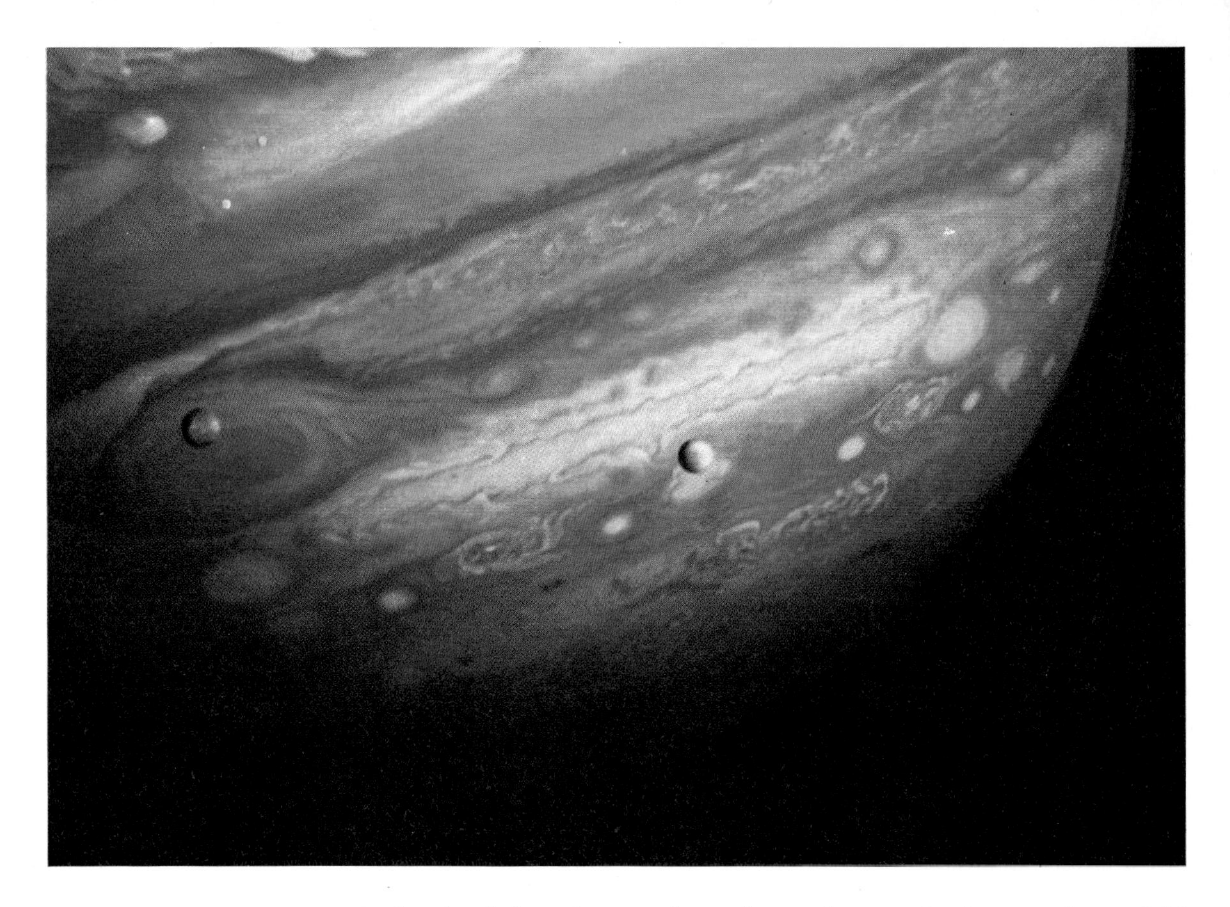

Jupiter; a Voyager 1 photograph taken on 13 February 1979 from a range of 12,400,000 miles. At this resolution (250 miles) there is definite evidence of circular motion in Jupiter's clouds. While the dominant large-scale motions are west-to-east, small-scale motions imply eddies within and between the belts. Two satellites are shown, Io (*left*) and Europa; the redness of Io is very evident even when seen against the background of the Great Red Spot.

Jovian day is less than ten hours long – but not in the way that a rigid body would do; the region near the equator moves fastest, and has a rotation period around five minutes shorter than that at higher latitudes. There are some features of particular note, such as the Great Red Spot, a huge oval marking that can sometimes be as much as 30,000 miles long by 9,000 miles wide, giving it a surface area greater than that of the Earth. There are four large satellites, known collectively as the Galileans because they were seen by Galileo Galilei as long ago as 1610, plus numerous smaller ones, most of which are too faint to be seen with ordinary telescopes.

Even before the Pioneers began their journeys we knew a good deal about Jupiter. In particular it is quite definitely a planet, not a junior star. A star shines by its own energy, which means that the internal temperature must rise to millions of degrees. Jupiter is certainly hot inside – the temperature may be at least 50,000 °F., perhaps more – but it is not nearly hot enough to trigger off stellar-type nuclear reactions.

Various theories of its internal structure had been proposed. The best of them assumed that there is a central rocky core, larger than the Earth and much more massive, surrounded by deep layers of liquid hydrogen, above which comes the gaseous 'atmosphere' that we can actually see. (With a world built on this pattern, it is not entirely easy to decide just where the 'atmosphere' ends and the planet proper begins.) It was also known that Jupiter is a source of radio waves, and it was assumed that there must be a powerful magnetic field. Yet there were many unknowns. For instance, why was the Great Red Spot red? And what about the other colours seen in the upper clouds? Moreover, even the Galilean satellites looked so small as seen from Earth that almost nothing was known about their surface features.

When the Pioneers set out, it was obvious that in some ways their missions were more difficult than anything previously attempted, if only because Jupiter is so far away that the outward journeys would take over a year and a half. If any serious fault developed, there would be little hope of putting it right. Moreover, solar power could not be used for the transmitters and other purposes, because in those desolate regions there is not enough sunlight; instead, a tiny nuclear generator was installed. Everything had to be as lightweight as possible, so that the number of experiments which could be carried out was strictly limited.

Another potential hazard was the asteroid belt, which lies between the orbits of Mars and Jupiter. Here we find thousands of midget worlds, together with some larger ones; Ceres, the senior member of the swarm, is over 600 miles across. The main asteroids were known, and could be avoided, but many of the smaller ones were not, and a collision between a spacecraft and a chunk of rock the size of a writing-desk could have only one result. There was no way of avoiding the asteroid belt; the Pioneers had to take their chance, and some of the technicians at J.P.L. were by no means confident of the outcome.

Luckily there were no hitches. Pioneer 10 by-passed Jupiter at 82,000 miles from the cloud-tops on 3 December 1973, and its twin made an even closer rendezvous almost exactly a year later, on 2 December 1974. The asteroid belt was safely negotiated; the instruments worked faultlessly, and the pictures received were better than anyone had dared to hope. Yet Pioneer 10 very nearly failed. Jupiter's radiation belts, of the same general type as our Van Allen zones, proved to be much stronger than had been anticipated, and the spacecraft's instruments were almost saturated. A few thousand miles closer in, and they would have been put permanently out of action. Luckily there was ample time to adjust the path of Pioneer 11 so that it passed quickly over Jupiter's equatorial region, where the radiation hazard was at its worst.

The temperature measurements were particularly significant. It was known that the cloud-tops are cold (around −180 °F.), but it was also known that Jupiter sends out more energy than it would do if it depended entirely upon what it receives from the Sun, and this was fully confirmed. More surprisingly, the equatorial and polar temperatures were found to be much the same, so that presumably the poles received more than their fair share of heat from below. The magnetosphere – that is to say, the region of space over which Jupiter's magnetic field is dominant – proved to be extremely large, extending out for millions of miles, and to be very complex. The magnetic poles are reversed relative to those of the Earth, so that if it were possible to use a magnetic compass there the needle would point south.

The colours of the belts and zones were striking, and the Great Red Spot was very much in evidence. Up to that time nobody had been sure of its nature; it had been regarded variously as a sort of floating island, the top of a column of stagnant gas, or (much less plausibly) a volcano-top. Pioneer showed that all these ideas are wrong. The Red Spot is a whirling storm, a phenomenon of Jovian 'weather'; presumably it has lasted for so long because it is exceptionally large, and therefore more persistent than smaller spots, although it may well die out in the foreseeable future. The top of the Spot is higher and colder than its surroundings, indicating that we are dealing with a high-pressure area. As it rotates it sucks in minor features, whirling them around, and altogether it affects the whole of that part of Jupiter.

Having made its pass, Pioneer 10 began a never-ending journey out of the Solar System. It will never come back; it is travelling too fast to be kept in captivity by the pull of the Sun, and it will become Man's first messenger into interstellar, as opposed to interplanetary, space. At the time of writing (February 1982), it has already crossed the orbits of the next two giant planets, Saturn and Uranus, and it is still sending back data. We may hope to keep in touch with it for several years yet, but then we will hear it no more; it will continue its journey between the stars, invisible and inaudible, until it collides with an object large enough to destroy it. This may not happen for millions or hundreds of millions of years, and it is even possible that it will outlast the Earth itself. In case it is ever found by some alien civilization, it carries a plaque which is intended to show its place of origin. (I can only comment that if the inhabitants of another world do find the plaque, they will need considerable intelligence if they are to interpret it!) The chances are slight, though I suppose that they are not nil.

Next came Pioneer 11, which confirmed all the earlier findings and sent back more high-quality pictures, showing definite changes in the cloud-layer. Unlike its predecessor, it had a further role to play. After the encounter with Jupiter there was sufficient power left for it to be swung round, back across the Solar System, on to a rendezvous with Saturn. This was achieved on 1 September 1979, and we were treated to our first close-range views of the wonderful Saturnian rings. The pictures were remarkably good, bearing in mind that Saturn had been merely an afterthought in the mission, and the encounter was doubly valuable because it showed that a spacecraft could go reasonably near Saturn's rings and emerge unscathed. Pioneer 11, too, is now on its way out of the Solar System – complete with plaque.

The scene was set for the next probes, Voyagers 1 and 2, which were even more ambitious. They were deliberately designed to study both Jupiter and Saturn, while Voyager 2 was also scheduled to go on to the outer planets, Uranus and Neptune, as well. This was possible because of an unusual positioning of the planets towards the end of the 1970s, not to recur for over a century and a half in the future. The four giants, Jupiter, Saturn, Uranus and Neptune, were strung out in a curve, so that they were all on the same side of the Sun.

The plan was to send the Voyagers first to Jupiter, and then use Jupiter's powerful gravity to swing them round and propel them on towards Saturn. Voyager 2 would then use Saturn's gravity to send it on its way to Uranus, and Uranus in turn would then send it out to Neptune. It was an extension of the two-planet method followed by Mariner 10

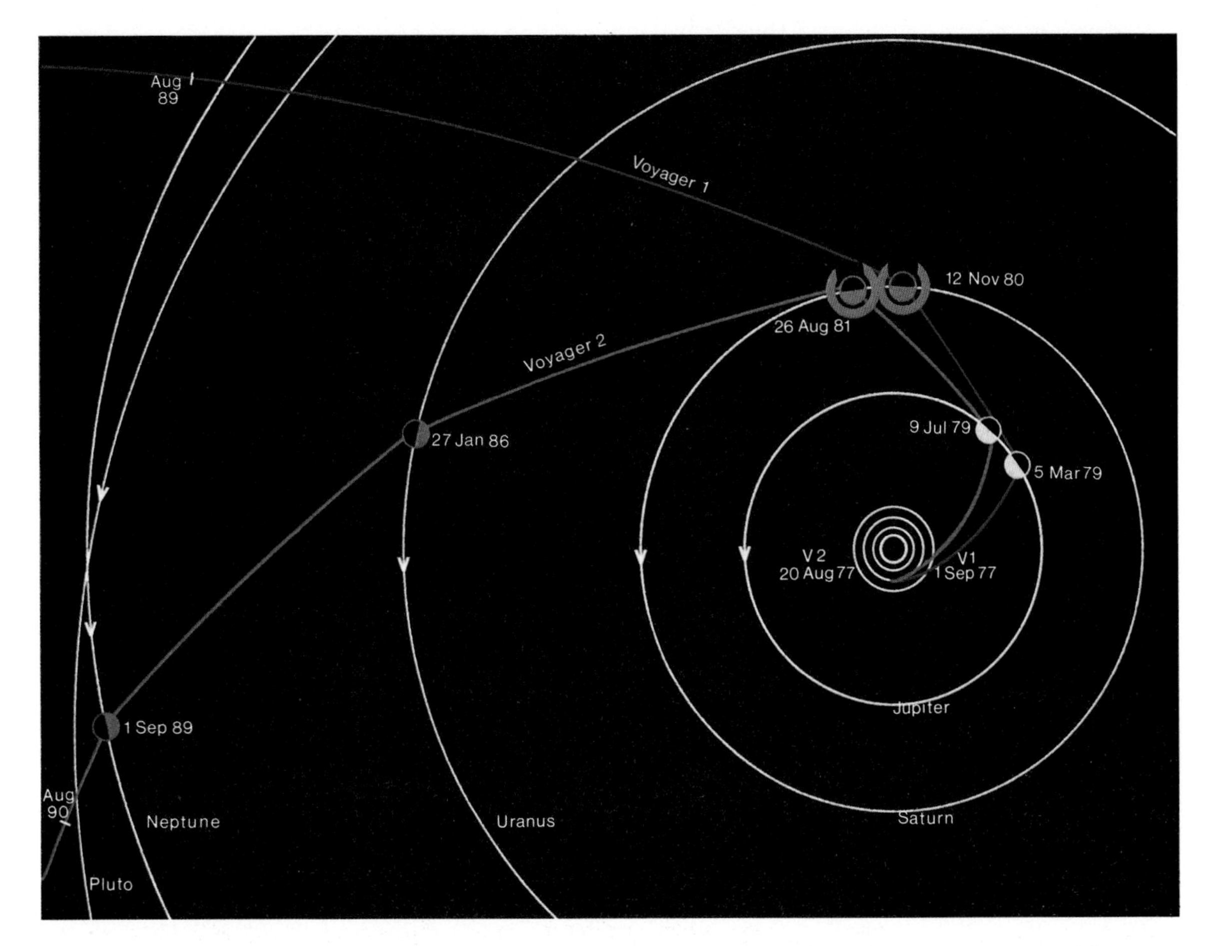


Voyager 1 and 2 trajectories, with launch and encounter dates. Also given are the approximate dates when the spacecraft will move out beyond the orbit of Pluto, thereby leaving the main part of the Solar System permanently. Diagram by Paul Doherty

in the inner part of the Solar System, though it was considerably more difficult.*

It may be asked why only Voyager 2 was scheduled to make the full 'Grand Tour'. The reason is that Saturn has a particularly interesting satellite, Titan, and for technical reasons it was impossible to survey both Titan and then Uranus and Neptune on the same mission. Therefore, Voyager 1 was programmed to by-pass Titan, and then go on its way to outer space; Voyager 2 would ignore Titan, and so would be able to move out towards Uranus and Neptune.

For once there were some preliminary hitches. The first Voyager gave trouble in the pre-launch period, and eventually it was switched with its companion, so that the original Voyager 1 became Voyager 2. To confuse matters still further, Number 2 was launched first, on 20 August 1977,

*This situation led to a curious episode which caused a good deal of alarm. A book called *The Jupiter Effect*, by Dr John Gribbin and Mr Stephen Plagemann, claimed that since all the planets would be pulling in the same direction during 1982 there would be cataclysmic results; the Sun would be affected, and this would in turn produce storms, cyclones, earthquakes and all manner of other disasters on Earth. The book was widely publicized, and things were not helped by sensational articles which treated it seriously. Of course it is absurd; even when added together the pulls of the planets could have no measurable effects at all – and, moreover, any elementary calculation was enough to show that there would not be even an approximate planetary alignment in 1982. Nevertheless, a scare had been started, and took some time to die away. Let us hope that we have now heard the last of the mythical Jupiter Effect!

leaving its twin to follow on 5 September. However, Voyager 1 was moving in a more economical transfer orbit. During the crossing of the asteroid belt it took the lead, and finally reached Jupiter on 5 March 1979, over four months before its companion.

This time the pictures were even better than those from the Pioneers. The vividly-coloured belts, zones and spots stood out magnificently, and new data were sent back, settling more of the outstanding problems. The cloud-layer was likened to a kettle of bubbling paints which refused to mix, and it became clear that Jupiter is a world of ceaseless activity; there is nothing calm or placid about it.

According to the latest information, Jupiter is indeed mainly liquid. Surrounding the rocky core the pressures are tremendous – thousands of times the pressure of our air at sea-level – and under these conditions hydrogen, which makes up about 80 per cent of Jupiter, becomes not only liquid, but also metallic. In this state it is a good conductor of electricity, and as it is stirred by the effects of the intense heat a strong magnetic field is generated. Above the metallic shell is a deep 'ocean' of more normal liquid hydrogen, above which come the clouds. Apparently there are three main cloud-layers. The lowest is composed of water ice or perhaps liquid water droplets; then come crystals of a chemical

Jupiter's faint ring, about 6 miles thick and 4000 miles wide, is shown in this colour composite as two light orange lines protruding outward to the right from Jupiter's limb. This photograph of Jupiter's dark side was taken through orange and violet filters while Voyager 2 was moving through the planet's shadow. The multiple images of Jupiter's bright limb are evidence of the motion of the spacecraft during these long exposures. The range was 900,000 miles, and Voyager 2 was about 2° below the ring plane. The upper ring image was cut short by Jupiter's shadow on the ring.

substance known as ammonium hydrosulphide, which is a compound of ammonia and the well-known, evil-smelling hydrogen sulphide ($H_2S$); and finally a cloud-layer made up of ammonia ice.

There are other compounds too, among which is phosphine. It is quite likely that when this phosphine rises to the top of the cloud-layer it is broken up, producing red phosphorus – and this may explain the colour of the Great Red Spot.

The radio results were also striking. Jupiter is a noisy planet; violent thunderstorms are raging all the time, and there is incessant lightning, while as the Voyagers passed around the night side of the planet they recorded auroræ, essentially the same as our polar lights but much more brilliant. Finally, Jupiter was found to have a ring, less than 30 miles thick, and too dim to be seen from Earth. It was discovered by Voyager 1, and the planners at NASA had ample time to adjust the path of Voyager 2 so as to study it as carefully as possible.

Seen in the spacecraft pictures the ring looks almost as though it is solid, but there is no chance that a solid or liquid ring could exist there; it would be promptly torn to pieces by Jupiter's gravitational pull. Instead, the ring is made up of very small particles which might almost be classed as 'dust'. It is quite different from the magnificent rings of Saturn, where the particles are made up of ice; and it may not even be a permanent feature, though it is certainly a long-lived one.

There was also the satellite system, and it was here that the most dramatic discoveries were made, not so much with the smaller members of the family but with the Galileans: Io, Europa, Ganymede and Callisto.

All are so large that they are of planetary size. Io is 2260 miles in diameter, Europa 1940, Ganymede 3280 and Callisto 3010, so that only Europa is smaller than our Moon, while Ganymede is actually larger than Mercury, though less massive. Before Voyager, it was tacitly assumed that all four would be of the same general type – but nothing could have been further from the truth.

Callisto, outermost of the Galileans, is also the least dense, and contains a great deal of ice, though there is probably a rocky core several hundred miles across. The immediate impression from looking at a Voyager picture of it is of craters, craters and still more craters; the whole area is saturated, and there seems to be practically no level surface at all. Some of the craters have central peaks, and some are at the core of bright ray systems. There are also two large circular structures, picturesquely named Valhalla and Asgard, with concentric rings around their brighter centres. Valhalla, the larger of the two, has a central region 370 miles across, with an outermost ring almost 1800 miles in diameter.

Unquestionably Callisto's surface is icy and ancient. It may in fact be the oldest landscape we have yet seen in the Solar System. Callisto is a dead world, and nothing can have happened there for thousands of millions of years.

Ganymede, slightly larger and considerably denser than Callisto, also has an icy crust, but here the ice-layer may not be more than 50 or 60 miles deep, and below it comes a mantle of water or soft ice surrounding a rocky core. Craters abound, but some regions are almost free of

Callisto, photographed on 7 July 1979 by Voyager 2 from 1,438,000 miles. Callisto is the most heavily cratered body yet studied.

them, and there are curious bright stripes and grooves together with reasonably high ridges. Crossing Ganymede would be rather like trekking over a crunchy glacier. There are several darkish circular areas, the largest of which has been appropriately christened Galileo Regio.

The main difference between the two outer satellites is that Ganymede shows traces of past crustal disturbances, so that its surface may be rather younger than that of Callisto; but again we are dealing with a dead world, and in spite of its size there is no trace of atmosphere.

As we draw inwards towards Jupiter we come next to the smallest and most reflective of the Galileans, Europa – and a completely different scene. Again the surface is icy, but there are practically no craters at all; neither are there mountains or valleys. There is a more or less complete absence of vertical relief, and it has been said that Europa is as smooth as a billiard-ball. All that can be seen are bright and dark lines, criss-crossing each other to make up a pattern as complicated as any maze. They may well be fissures in the surface ice which have been filled by water or softer ice welling up from below, but they are very shallow indeed, and there is nothing orderly about them.

In its own peculiar way, Europa is as perplexing a world as we have yet encountered, and the absence of craters has to be accounted for in some way. If the craters on Ganymede and Callisto really were caused by meteoric bombardment, as many people assume, then how has Europa

Ganymede, from Voyager 1; March 4, 1979, from 1,600,000 miles. The several dots of a single colour (blue, green and orange) on the disk are the result of markings on the camera used for positional measurements, and are not physical features on Ganymede.

escaped so completely? All we can do is to say that the original impact craters have been wiped out by water or soft ice seeping out from below, but this is not a satisfactory solution, and it adds to my personal doubts as to whether we are thinking along the right lines. Internal forces also may produce craters, and in this case the smoothness of Europa would presumably be connected with its lesser distance from Jupiter.

Io, still closer-in, is different again, and spectacular by any standards. The surface is brilliant red, and the general appearance has been compared with that of a pizza. No impact craters exist – and none can be expected, because Io is in a state of constant turmoil. Live volcanoes hurl material high above the surface, making Io the most active world in the entire Solar System.

The initial discovery was made by Linda Morabito, one of the scientists at J.P.L., who was studying the Voyager results as they came in. On 9 March 1979 she was checking on the position of Io relative to a faint star when she saw a huge, umbrella-shaped plume rising from the edge of Io itself – and this could only be an active volcano, sending a plume to a height of over 170 miles.

Before long other eruptions were identified, some of them more violent than anything which could be produced by our Vesuvius or Etna. One of the most striking of the volcanoes, now called Pele, was seen to be heart-shaped, with a diameter of 750 miles; it ejected material

Europa, from Voyager 2; range 150,000 miles. Europa was not well shown from Voyager 1, and all that could really be said was that it was quite unlike the other Galilean satellites. It was not until the Voyager 2 results that the extent of the dissimilarity was fully appreciated. Maps of Europa are very difficult to compile, and no particular feature stands out prominently. It may even be said that one part of Europa looks very much like another!

at speeds of up to 2200 m.p.h. Others were equally explosive. Eight volcanoes were identified on the Voyager 1 pictures; when Voyager 2 flew past, four months later, six of them were still erupting, while two new ones had started. There were blackish spots which appeared to be crusted-over lakes of lava-like sulphur, with temperatures of over 80 °F., and it is possible that the main eruptions would send a thermometer up to at least 800 °F., though the normal surface has a day-time temperature of more than 200 °F. below freezing-point.

Sulphur is the key to the situation. It covers Io, and it may well be that the thin crust overlies a deep ocean of liquid sulphur surrounding the rocky core. If it were possible to go there, the scene would be weird in the extreme, with pastel yellow, orange or blue 'snow' raining down from the sky together with blobs of molten sulphur. Near the Ionian equator there could be white patches of sulphur dioxide frost, and crusted-over black lakes, caused by floods of molten black sulphur flowing out through fissures and then freezing. Io would be a world to avoid even if it were not totally immersed in the lethal radiation zones surrounding Jupiter.

In fact Io was already known to have marked effects on the Jovian radio waves, and apparently there is an electric current flowing between the two bodies with a strength of about 5,000,000 amperes – equal to all the Earth's power plants combined. Jupiter's radiation bombards the Ionian surface, sputtering off a cloud of tenuous material; it is even possible that material from Io cascades down into the Jovian atmosphere to produce the brilliant auroræ recorded by the Voyagers. The intense activity on Io is thought to be due to the periodical flexing and churning of the whole globe by tidal interactions both with Jupiter and with Europa,though we cannot yet decide how Io 'works'.

Still closer to Jupiter are several junior satellites. One of them, Amalthea, was discovered as long ago as 1892, and was shown by Voyager to be irregular and red, with two features which look as though they could be mountains, but its longest diameter is less than 200 miles, and it is unlikely that there are any active volcanoes there. Well beyond the Galileans are further satellites, probably captured asteroids; the outer four move round Jupiter in a wrong-way or retrograde direction, rather like cars that have turned the wrong way on a roundabout.

The Voyagers have long since left Jupiter behind, and as yet only one more probe is planned there. This is Project Galileo, due to be launched in April 1985. It will reach its target in mid-1990, and the orbiting section will fly past Io at 600 miles before entering a closed path round Jupiter itself. The 'entry' vehicle will then be separated, plunging into the Jovian clouds at over 100,000 m.p.h. It will soon be slowed down, partly by friction and partly by a deployed parachute, and 40 minutes later it should have sunk below what are thought to be Jupiter's lowest water clouds. As it sinks deeper and deeper it will have to endure increasing pressures and temperatures; it can hardly survive for longer than an hour at most before being destroyed, but it should be able to send back important new information, and the orbiter may be expected to go on transmitting for several months at least.

Meanwhile, what about Saturn, next on the list of Voyager targets? Here we have a world which is in some ways similar to Jupiter and in others markedly different. The general pattern is the same; a solid, rocky

Io, from 300,000 miles. The plume at the limb rises from one of the most violent of the Ionian volcanoes, Loki, which was erupting during both the Voyager 1 and Voyager 2 passes, though its shape had altered perceptibly. The position of Loki on Io is latitude 19 °N., longitude 305 °W. Another major volcano, Pele (latitude 19 °S., longitude 257 °W.) was just as violent as Loki during the Voyager 1 pass, but was not erupting when Voyager 2 made its fly-by, though it is very unlikely to have become extinct.

core, relatively larger than Jupiter's, overlaid by liquid hydrogen, the lower layers of which are metallic, together with a good deal of helium. The general aspect of the upper cloud-layer is blander than Jupiter's, because there is a greater amount of haze, and the belts are less prominent. Neither is there anything as conspicuous as Jupiter's Great Red Spot. The rotation period is 10 hours 39 minutes, while the Saturnian year is $29\frac{1}{2}$ times as long as ours.

Like Jupiter, Saturn radiates more energy than it receives from the Sun but for a different reason. The internal heat of Jupiter is simply left over, so to speak, from the time when the planet was formed nearly 5000 million years ago, but Saturn has had quite enough time to lose all its original heat, so that the extra radiation must be due to another cause. Probably the clue is to be found in the amount of helium mixed with the hydrogen. Saturn has cooled down so much that near the top of the gas-layers the helium has condensed to form liquid drops, and is now falling slowly towards the core, releasing energy as it rubs against the hydrogen 'fluid' through which it has to pass.

The glory of Saturn lies in its rings, which make, in my view at least, the most wonderful spectacle in the sky. There are three main rings, two (A and B) bright and the third (C, or the Crêpe ring) closer in and semi-transparent. Rings A and B are separated by a gap known as the Cassini

The clouds of Saturn, photographed by Voyager 2 on 19 August 1981 from 4,400,000 miles. The colours have been enhanced and in some cases modified to make scientific analysis easier; thus the blue patch, associated with obvious atmospheric movements, should really be brown.

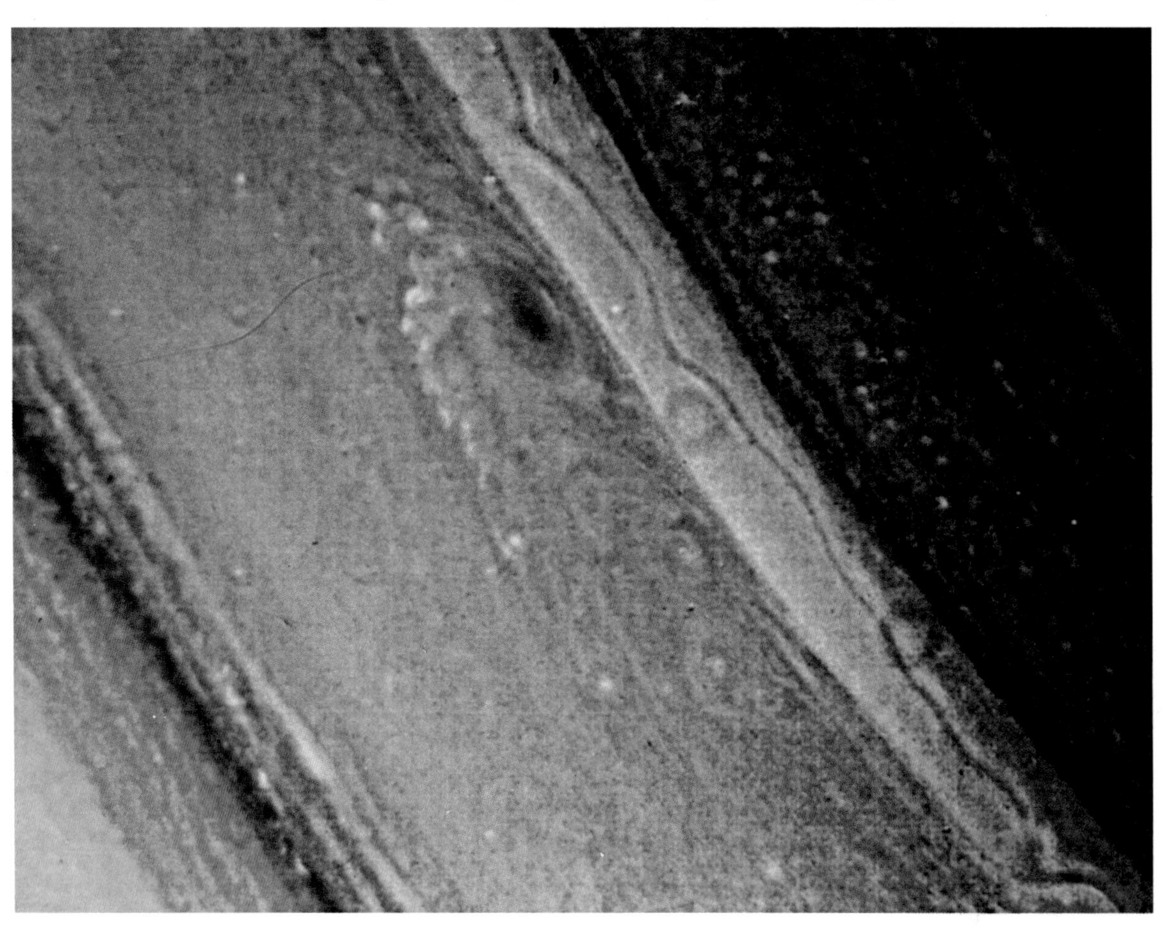

Saturn from Voyager 2; range 13,000,000 miles; date 4 August 1981. There are clear traces of the radial 'spokes' crossing Ring B. The three satellites shown are Tethys, Dione and Rhea.

OVERLEAF: An impression of Saturn's rings, which are made up of particles of ice, rock and dust ranging from large blocks to tiny grains. In this view we are assumed to be observing from one of the sparsely populated regions of the rings, looking in to the denser parts of the rings towards the planet. Saturn itself forms a colourful backcloth to the ring fragments. Painting by Paul Doherty

Division in honour of its discoverer, while in Ring A there is a less prominent gap, Encke's Division. The rings measure 169,000 miles from one side to the other, but they are also very thin, so that when viewed edgewise-on they almost disappear, as last happened in 1980. (Using the Lowell, 24-inch refractor I was able to keep them in view, but only with great difficulty.) The rings are made of ice particles, ranging in size from marbles to blocks several feet across.

Our first close-range view of the rings was obtained from Pioneer 11, and an extra outer ring was discovered, but Pioneer was not originally programmed to study Saturn, whereas the Voyagers were. Voyager 1 made its pass on 12 November 1980, and, as before, the Press Room at J.P.L. was crowded with newsmen from almost every country in the world, including the U.S.S.R. We expected something really spectacular, and we were not disappointed.

As Voyager homed in towards Saturn, details on the disk started to show up. As with Jupiter there were obvious belts, together with spots – one of them distinctly reddish – and there was evidence of violent activity; it was subsequently found that the winds on Saturn are the strongest known in the Solar System, attaining velocities of well over 1000 m.p.h. There are marked differences between Saturn's general

The rings of Saturn, as seen from Voyager 2. The colour variations indicate differences in chemical composition. Colours shown here have been enhanced and modified to facilitate analysis.

atmospheric circulation and that of Jupiter, and there is less to be seen, but there were brightish colours here and there.

In passing, it is worth noting that a 'colour picture' sent back from a spacecraft such as Voyager is not of the same type as an ordinary colour film. The pictures actually taken are black and white, but filters are used, and these allow the pictures to be reconstructed faithfully, though very often the differences in colour are deliberately exaggerated to make them clearer for scientific analysis. The procedure is highly complicated. When the signals first come in they show comparatively little, and it takes time, patience and skill to draw out all the details which are really there.

New data came in to J.P.L. all the time. Yes, Saturn had Van Allen zones, less dangerous than Jupiter's but still very strong; there was an extensive magnetosphere, and the magnetic field was much more powerful than that of the Earth. Radio emissions were checked, and the rotation period re-measured. But it was the rings on which the main attention was focused, and before long it became clear that they were not nearly so simple in structure as had been expected.

Instead of an almost featureless surface, the rings showed curious grooves. As each hour passed this became more and more evident, until there could be no doubt that each known ring was divided into hundreds of ringlets and gaps – it was even said at the time that the ring system had more grooves than any gramophone record. There were ringlets also in the Cassini and Encke Divisions, both of which had been expected to be empty; at one time there had even been talk of sending Pioneer 11 right through the Cassini Division.

This was only the beginning. Strange spoke-like markings appeared in the B ring, at distances of between 27,000 miles and 35,000 miles above Saturn's cloud-tops. They first came into view as they emerged from the shadow of the planet, and survived for several hours, becoming distorted as they swung round; when they vanished they were replaced by new spokes coming out of the shadow. This seemed apparently impossible, because ring particles do not all move at the same speed; those closer to Saturn travel faster than those further out, and no radial features ought to be formed. Yet there they were, and no obvious explanation was forthcoming. My own suggestion, made at the time, was that they were due to tiny particles elevated from the main ring plane by electrostatic or magnetic forces, but this may or may not be correct.

Earth-based observers had previously reported another ring (Ring D), closer to the planet than any of the others. I had always been sceptical about it, because I had never been able to see it even with large telescopes, and in a way I was right; Ring D does exist, and the Voyagers showed it, but it is poorly defined, and is not prominent enough to be visible except from a spacecraft. On the other hand Ring F, outside the main system, was amply confirmed, and was apparently clumpy; there were extraordinary braids, and it looked as though two or even three rings were intertwined, which again appeared to be dynamically impossible. The F ring is flanked by two tiny satellites, one to each side, which act as 'shepherds', keeping the ring particles firmly in their set orbits. If a particle strays, the shepherds will force it back again.

Outside the Ring F were two more, Rings G and E (by now the alphabetical sequence had become decidedly chaotic). And to complete the picture, there were several ringlets, including at least one in the Cassini Division, which were not perfectly circular.

All these revelations led to endless discussions, and there were equally fascinating discoveries in connection with the satellites. Nine had been certainly known before the Voyager flights, but the total mounted quickly, and two of the newcomers proved to be very unusual indeed. They move round Saturn beyond the Ring F, and share the same orbit – or practically so. Both are irregular in shape, with longer diameters of between 50 and 60 miles. At the time of the Voyager pass the smaller of the two was marginally the closer-in, and so slightly the faster-moving, but calculations show that every few years the two approach each other within a few miles, and then actually exchange orbits, a state of affairs unique in the Solar System. Probably they are the fragments of a single body which met with some mishap long ago.

Two of the familiar satellites, Dione and Tethys (both discovered by Cassini in 1684) proved to be sharing their orbits with smaller bodies. Dione has a companion moving well ahead of it, and Tethys two, one ahead and one behind. Since the orbits are the same, there is no danger of the dwarfs being swallowed up; they keep prudently well away from their larger neighbours.

Close-range views were obtained of several of the major satellites, but little was seen on the most important of them all, Titan, which has a diameter of not much less than 3000 miles. This was because Titan, alone of planetary satellites, has an atmosphere. The existence of this atmosphere had been known ever since 1944, when Gerard Kuiper detected it spectroscopically, but when Voyager began its journey nobody was certain whether or not the surface would be hidden by clouds, and opinions differed. I thought that we would probably see details; Garry Hunt, one of the Principal Scientific Investigators, did not – and in the event Garry was right. The atmosphere was dense enough to hide the surface completely, and Titan looked simply like an orange blob. Nevertheless, a good deal of information could be collected, and the project controllers heaved sighs of relief. Had Voyager 1 failed to inspect Titan, then Voyager 2 would have had to do so, and would have lost the chance of going on to Uranus and Neptune.

Of the named satellites, known before the space missions, three – Mimas, Dione and Rhea – were well surveyed by Voyager 1. All have icy surfaces. Rhea, over 900 miles in diameter, is heavily cratered, and Dione too has craters, together with bright, wispy frost patterns due probably to slushy material that has seeped out from below the crust. On Mimas there is one huge crater, now named Herschel, which has one-third the diameter of Mimas itself. Mimas is a mere 240 miles across, and Herschel measures 80 miles; if the crater had been formed by an impact Mimas would surely have been in grave danger of being shattered, particularly as its overall density is not much greater than that of ice.

Voyager 1 had done its work well, and we bade it farewell as it drew away from Saturn. It would go on transmitting for years to come, perhaps well into the 1990s, but it would encounter no more planets, until perhaps it entered some other solar system many light-years away. Meanwhile, which were the most important of its discoveries?

Despite Saturn itself, and the rings, I tend to regard Titan as being pre-eminent. True, nothing could be seen of the surface, but enough data had been collected to show that once again we have a world which is unique. The atmosphere is denser than ours, with a ground pressure one and a half times as great, and the main constituent is nitrogen, together with large quantities of methane and smaller amounts of other gases. Organic compounds are produced there by the action of sunlight, and there seem to be all the ingredients for life. The problem is that the low temperature, around $-288°$F., makes these ingredients form an icy sludge, stopping chemical action before life can appear.

Significantly, the surface temperature is within four degrees of what is termed the triple point of methane, i.e. the temperature at which methane can co-exist as a liquid, a gas and a solid, just as $H_2O$ can do on Earth. On Titan, then, there may be rivers and lakes of liquid methane, together with solid methane-ice cliffs and a steady methane rain falling from the nitrogen-rich clouds above. There may even be lakes of liquid nitrogen, and it was facetiously suggested that any future explorers will have to use a submarine rather than a more conventional type of probe!

But will a landing there ever be possible? There seems to be no reason why not. The actual surface is solid, and Titan is well beyond the dangerous part of Saturn's radiation zone, since it moves at the edge of the planet's magnetosphere. Unfortunately there is not much more to be found out at the moment. The only way to tackle Titan is to put a radar-equipped probe in orbit around it, as has already been done with Venus. Funds for this sort of experiment are not available as yet, dearly though scientists would like to see it carried out.

The Press departed; the project scientists settled down to their analyses, and J.P.L. reverted to its normal calm, but not for very long. Voyager 2 was approaching Saturn, and made its pass on 27 August 1981, going much closer-in than its twin had done, and returning pictures of even better quality. Again the multitude of ringlets came into view; the outer F ring was no longer braided, but the radial spokes were still there, persisting for hours before losing their identities.

Voyager 2 showed that the individual ringlets number not hundreds, but thousands. Though we cannot yet pretend to know the full solution, it looks very much as though we are dealing with a wave pattern of some sort, and it may well be that the larger satellites outside the ring system produce spiral pressure-waves, much as a ship will produce a

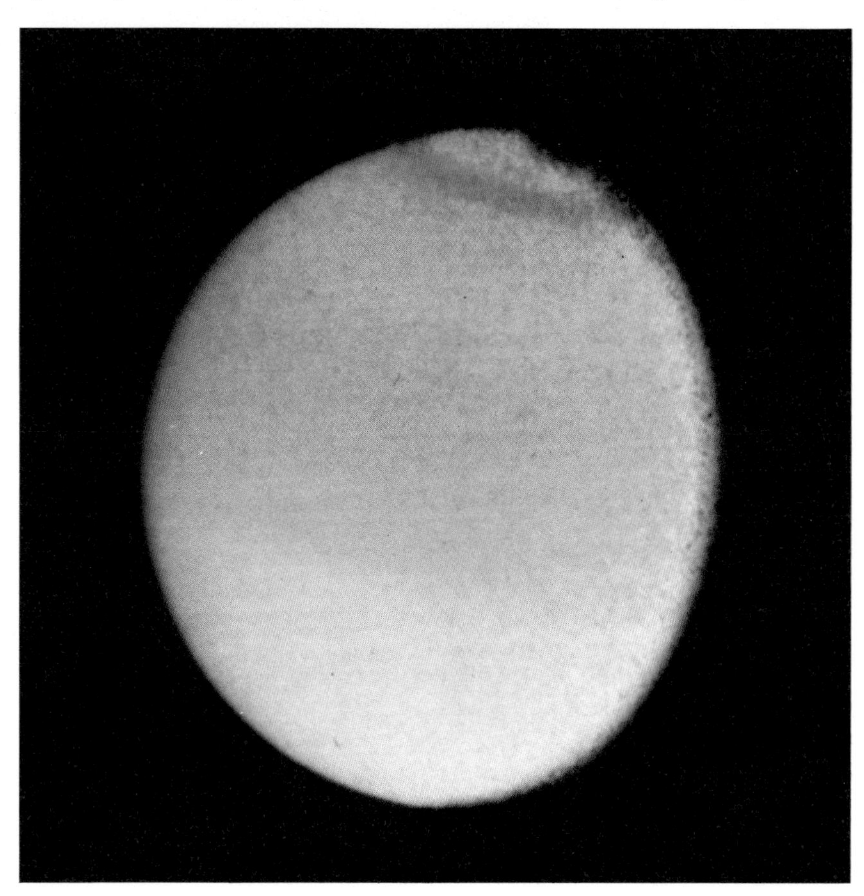

Titan, from Voyager 1; range 1,400,000 miles. No surface details are visible, but the colour is not uniform; the two hemispheres differ, possibly because of seasonal effects (remembering that Titan shares in Saturn's motion round the Sun; the revolution period is $29\frac{1}{2}$ Earth years).

Iapetus; Voyager 2 photograph from 680,000 miles, showing craters over both the bright and the dark parts of the satellite.

wake as it travels across a calm sea. The small-scale structure of the rings may even be very variable, though the main gaps, such as the Cassini and Encke Divisions, are presumably permanent.

Voyager 2 also surveyed the main satellites which had not been satisfactorily covered from Voyager 1; in order of decreasing distance from Saturn, Phœbe, Iapetus, Hyperion, Tethys and Enceladus. Phœbe, over 8,000,000 miles out and a mere 100 miles across, was found to be reddish and relatively dense. It is exceptional inasmuch as its rotation is not synchronous; that is to say, it does not always keep the same hemisphere turned towards Saturn, as the other satellites do. Phœbe's orbital period is 550.4 Earth-days, but it spins round in only nine to ten hours. Iapetus, moving at a distance of more than 2,000,000 miles from Saturn, does have a synchronous rotation (79 Earth-days), and is a larger body, some 900 miles in diameter. It is interesting because one of its hemispheres is as bright as ice and the other as dark as a blackboard. This had already been known, because of its regular fluctuations in brightness as seen from Earth, but the cause was obscure – and for that matter it still is. Because the density of the globe is so low, it seems safe to assume that it is the dark part which is the deposit, and there have been suggestions that material has been sputtered off Phœbe and wafted in to cover the leading side of Iapetus, though personally I am doubtful, because Phœbe is very small, and never comes within 6,000,000 miles of Iapetus.

More probably the dark coating has welled up from inside Iapetus itself, though in this case it is not easy to tell why none of the other satellites are similarly divided. Both the dark and the bright areas contain craters, some of them large.

Hyperion also presents problems of its own. It is irregular in shape, and has been likened variously to a lozenge, a potato and a hamburger; the maximum diameter is 225 miles. There are plenty of craters, but the oddest feature is that Hyperion does not point its longest diameter towards Saturn, as dynamically it ought to do. Even if it had been knocked out of alignment by a violent impact it might be expected to take up its stable attitude eventually, though there is always the chance that it has not yet had time to do so. It gives the impression of being a fragment of a larger body, but, as I asked plaintively – where's the other half?

Tethys, 650 miles across and much closer to Saturn, is little if at all denser than water, and presumably consists of almost pure ice. Here we have a truly enormous crater, larger than the whole body of Mimas, and also a gigantic fracture, now named Ithaca Chasma, 460 miles long and nearly 40 miles wide, together with assorted craters and other minor features. Tethys is exceptionally cold; the surface temperature was given as –305 °F.

Next comes a smaller satellite, Enceladus (diameter 310 miles) which is equally exceptional. Part of the surface shows relatively small, young-

Tethys; Voyager 2, range 125,000 miles: date 26 August 1981. Ithaca Chasma, the huge trench girdling about three-quarters of the circumference, is well seen.

Enceladus; Voyager 2, from 74,000 miles; date 25 August 1981. The two different types of terrain are well shown. (At the time it was commented, rather acidly, that Enceladus and Europa have several things in common; both are the second of the long-known named satellites in order from their primaries; both are unlike the other members of their respective families in being comparatively smooth; and both have names beginning with E!)

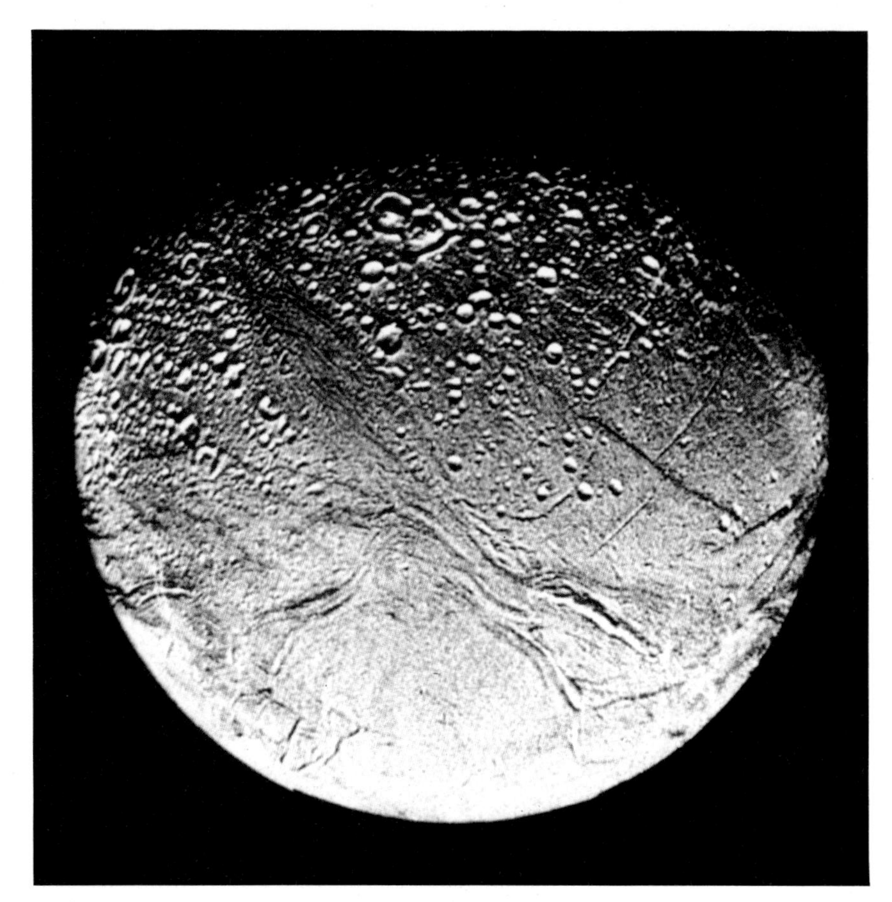

looking craters, while another part is crater-free, with nothing more than stripes and grooves. Evidently there has been past internal activity, so that water has flooded out from below the crust and obliterated any existing large craters, so playing the same role as lava has done on the Moon. Tidal forces are presumably responsible, but it is difficult to explain why Enceladus is so unlike its two cratered neighbours, Mimas on the inner side of its orbit and Tethys further out. Incidentally, Enceladus seems to be the most reflective body in the whole of the Solar System.

Up to this point everything had gone well. The next part of the programme was to take place as Voyager 2 drew away from Saturn, *en route* for Uranus. One experiment was particularly ingenious. As the spacecraft flew by above the rings, it could study a bright star, Delta Scorpii, which lay on the far side; the star would be seen through the gaps, and would be hidden by the ringlets themselves. The experiment worked to perfection, showing that there were thousands of ringlets and that there are very few completely empty gaps anywhere. But there was trouble ahead. As it passed through the ring plane on its outward journey, the spacecraft was apparently struck by an ice particle, and the platform carrying the main cameras and other instruments jammed. When ordered to rotate, it stubbornly refused to do so.

This could have wrecked not only the final part of the Saturn encounter, but also the missions to Uranus and Neptune, because if the scan platform could not be accurately pointed it would be unable to send back any data. The problem arose when Voyager was behind Saturn, and it caused consternation in Pasadena; it is rather difficult to carry out running repairs from a distance of over a thousand million miles, and the only course was to play back the tape recordings which had been received, study them from the engineering point of view, and then transmit what seemed to be the most suitable command.

This was accordingly done, and the platform obeyed its orders – more or less; but the problem was by no means completely resolved, and at the moment, with Saturn left far behind and Uranus still far ahead, we can only hope that the platform will function at least adequately when the need arises. During the long glide out to Uranus the spacecraft is merely sending back routine data about conditions in that part of the Solar System, and to lose these is of comparatively minor importance, so that the scan platform will be moved as little as possible until Uranus is close ahead.

By 1985, when the Uranus encounter is due, we will know the result. If the cameras can be used to their full capacity, both with Uranus and later with Neptune, Voyager 2 will be accounted 100 per cent success. Even if not, it must still be counted as one of the major triumphs of the Space Age to date.

# 8 The Edge of the Solar System

Of all the planets it is perhaps Uranus that provides the closest link between the present and the past. It is too dim to be seen with the naked eye, unless you know exactly where to look for it, and it was not discovered until the year 1781, when it was found by the man who was possibly the greatest observer in astronomical history: William Herschel, Hanoverian by birth, English by adoption. Herschel built his own reflecting telescopes, and used them to carry out what he termed 'reviews of the heavens'. It was in this way that he identified Uranus, which showed a disk and was obviously not a star.

Herschel made his discovery from the garden of a house in the city of Bath, Number 19 New King Street, which has recently been renovated and turned into a Herschel museum. (Should you be in that part of England, I suggest that you pay it a visit.) The telescope he was using had a mirror only six inches in diameter, but it was extremely good, and Herschel used it well. His planet was named Uranus; it proved to be a giant, over 30,000 miles in diameter, moving round the Sun in a period of 84 years. Neither Herschel's telescope nor anyone else's could show definite details on the pale, greenish disk, so that until the last 15 years or so we knew comparatively little about it, and it was only in 1977 that there were really major developments.

Our ignorance was not complete. Spectroscopes confirmed that Uranus has a gaseous surface, and theorists concluded that there must be a solid core, overlaid by the usual layers of liquid hydrogen, which are in turn covered by the atmosphere. Yet Uranus is not merely a smaller edition of Jupiter or Saturn. There is less hydrogen and helium, and much more ammonia, methane and $H_2O$. At its tremendous distance from the Sun (1783 million miles on average) Uranus is a very chilly place indeed, and it seems to have no internal heat-source. The outer atmosphere is not only very cold, but very clear to great depths.

A curious feature of Uranus is that its axis is tilted at more than a right angle. On Earth, the angle is $23\frac{1}{2}$ degrees, which is why we have our familiar seasons, but the value for Uranus is 98 degrees, and the calendar there is truly remarkable. Each pole has a night lasting for 21 Earth-years, with a corresponding midnight sun at the opposite pole; half an Uranian year later the conditions are reversed. Sometimes one pole faces the Sun, so that as seen from Earth it lies in the middle of the disk – as will be the case in 1985. Nobody has yet come up with a really good explanation of why Uranus is tilted to such an extent.

Artist's impression of Pluto and Charon. The surfaces of the two bodies are only 12,000 miles apart, and Charon would be an impressive object in Pluto's sky. The Sun is so distant that it appears only as a very brilliant star, casting a relatively feeble light across the frozen Plutonian landscape. Painting by Paul Doherty

Herschel's 40-foot reflector. (The total length was 40 feet; the diameter of the main mirror was 49 inches.) The telescope, brought into full use in 1789, was a cumbersome arrangement; it was difficult to operate, and the observer had to communicate with his assistants by means of a speaking tube. The optical system was of the front-view type, so that there was no secondary mirror; the main mirror was tilted so as to produce a focus at the edge of the top of the tube. The observer's platform is shown just below the tube-top. This was much the largest telescope ever built up to that time, and it was not surpassed in size until 1845, when Lord Rosse completed his 72-inch reflector at Birr Castle in Ireland. Using the 40-foot reflector, Herschel discovered two satellites of Saturn (Mimas and Enceladus), but it cannot really be said that the telescope was a success, mainly because of its awkwardness, so that most of Herschel's work was carried out with smaller instruments. The telescope was dismantled before Herschel died; the mirror is now on view at Flamsteed House (the old Royal Observatory at Greenwich) together with all that remains of the tube.

OPPOSITE: The main entrance to Herstmonceux Castle, present home of the Royal Greenwich Observatory. I took this picture in February 1982, when we were filming there for the 25th anniversary of the 'Sky at Night' television series; some of the BBC team can be seen on the drawbridge over the moat.

The first serious efforts to record details on Uranus were made in 1970, when three astronomers from Princeton University in the United States went up in a balloon, taking a telescope and photographic equipment with them. They reached almost 15 miles above sea-level, where the air is very thin and transparent, and took a number of pictures. Very little was shown, and in fact one has to use one's imagination to 'see' any detail at all.

Another part of the balloon programme was to measure the diameter of Uranus, which is more difficult than might be thought simply because the disk looks so small. The Princeton team gave a value of 32,200 miles, which may be a slight underestimate in view of very recent results. There was, however, another method of tackling this problem. A new technique was devised at the Royal Greenwich Observatory, Herstmonceux.

The Royal Observatory was founded in 1675, by express order of that much-maligned monarch Charles II, and was set up in Greenwich Park, where it remained until the 1950s. Unfortunately it had by then become more or less useless as an observing site, because of the spread of London, with its lights, its smog and its grime; so in a mammoth operation extending over many years, the working instruments were transferred to Herstmonceux in Sussex, leaving the old Greenwich buildings to be

turned into a museum. (It is, incidentally, one of the best in the world, and if you want to look back over the history of astronomy Greenwich is the place to visit.) The mathematicians also departed for Herstmonceux, and one of them, Gordon Taylor, became very interested in planetary and asteroid diameters.

Taylor made use of stellar occultations – that is to say, the times when stars are hidden by moving planets. Since the distance of the planet is known, its apparent diameter can be calculated from the length of the occultation, and its real diameter can then be found. The only drawback is that occultations by planets are not common, and one has to be in exactly the right place at exactly the right time; Uranus, with its tiny disk and slow movement, is not co-operative.

Taylor made careful calculations, and found that there would be an occultation on 10 March 1977, when Uranus would pass in front of a star known by its catalogue number of SAO 158687. The star was well below naked-eye visibility, and to make accurate timings of the moments of disappearance and reappearance would need electronic equipment used together with a relatively powerful telescope. Neither would the occultation be visible from Europe or the United States; it would be seen only from the southern hemisphere, and the exact path of the track over the Earth was uncertain within fairly wide limits.

Taylor alerted his southern colleagues. Close watches were organized in Cape Town and in Perth, but the best hope seemed to come from a most unusual source, the KAO or Kuiper Airborne Observatory.

This Observatory, named in honour of Gerard Kuiper, is nothing more nor less than a modified Lockheed C-141 aircraft. It carries a 36-inch reflecting telescope, installed in an open cavity recessed into the port side of the aircraft, immediately ahead of the wing. The aircraft can fly miles above ground level, and is remarkably steady; the telescope can be moved over a range of elevation from 35 to 75 degrees, and is capable of tracking celestial objects to an accuracy of less than two seconds of arc. At its peak altitude it is above 85 per cent of the Earth's air, and, of course, it can go more or less where it likes. To observe the occultation of SAO 158687, it was scheduled to fly over the Indian Ocean at a height of approximately 41,000 feet.

From the KAO, J. Elliot, T. Dunham and D. Mink successfully observed the occultation, which began at 20.52 hours G.M.T. and lasted for 25 minutes. But the main surprise came before the actual event. The star winked several times, as though it were being briefly covered up by material surrounding Uranus. Signals were promptly sent out by radio; could it be possible that Uranus had a system of rings? After the star reappeared from behind the planet there were more winks, and the two sets – pre-immersion and post-emersion – were symmetrical, which seemed to clinch the matter. The second set of winks was also observed by astronomers at Cape Town.

The discovery was quite unexpected. Saturn was regarded as unique in being attended by a ring system (the ring round Jupiter, remember, was not discovered until 1979, when Voyager made its pass), and nothing of the sort had been seriously considered with Uranus; but there was no doubt about it, and later observations have provided full con-

Uranus and the star SAO 158687, drawn by Paul Doherty on 11 March 1977 with the 16-inch reflector at his private observatory at Stoke-on-Trent. Magnification 480. This was the time when the Uranian rings were discovered, though from Stoke neither the planet nor the rings occulted SAO 158687. (My attempts to observe the phenomenon were unsuccessful. From my observatory, at Selsey, the sky was cloudy throughout the nights of March 10 and 11.)

firmation. During the 1978–80 period, five more occultations were observed, and each time the effects of the rings were seen. The first confirmation came on 10 April 1978, from Chile, where several major telescopes have been set up, notably the 100-inch Irénée du Pont reflector at Las Campanas, and the even larger instruments at the observatories of La Silla and Cerro Tololo. More observations came in 1980 from Sutherland, north of Cape Town, where the main South African telescopes have been concentrated. The final proof was obtained from Palomar, where infra-red techniques were used with the 200-inch reflector and the rings were actually recorded, though admittedly not very clearly.

By now we have found out a good deal about the Uranian rings. They are quite unlike those of Saturn. To begin with, there are at least nine of them, not all of which are perfectly circular. They are narrow; only a few miles broad in most cases. And, more significantly, they are not bright. The particles making up Saturn's rings are icy, but those of Uranus are as black as coal-dust.

Ground-based observations can do little more to improve our knowledge of this curious world, but if all goes well we may know much more within a few years. In 1985 the Space Telescope is due to be launched; it will be a 94-inch reflector, taken up in the Shuttle and put into orbit round the Earth. Operating from above the atmosphere it should be much more effective than any telescope now in use, and it should be capable of showing the Uranian rings. However, our main hopes lie with Voyager 2.

OVERLEAF: Impression of Uranus and its satellite Umbriel. In 1986 Voyager 2 will encounter Uranus; until then we can only imagine what the planet looks like from close range. There is a greenish atmosphere, and probably little detail. The satellites are believed to be made up chiefly of ice, and are no doubt cratered; whether or not Umbriel will show craters arranged in the manner shown here remains to be seen. The Uranian rings are much less spectacular than Saturn's, and at this immense distance from the Sun will appear dim and ghostly. Painting by Paul Doherty

I have already said a great deal about Voyager 2, and it is only necessary to add that the pass of Uranus is scheduled for 24 January 1986. The probe will send back close-range pictures; it will make temperature measurements; it will study the magnetic field and possible radiation zones of Uranus, and it will also pass within reasonably close range of all the five known satellites, Miranda, Ariel, Umbriel, Titania and Oberon, all of which are smaller than our Moon and are thought to be composed chiefly of ice. There is every reason to expect that they will be cratered, though it is unwise to be too dogmatic (remember Europa and Enceladus!). But, of course, everything depends upon that awkward scan platform, since if the cameras cannot be properly manœuvred, much of the value of the encounter will be lost (official estimates put the chances of success at 60 per cent).

Once Uranus has been left behind, Voyager 2 will proceed to its next and final target: Neptune, the outermost of the giant planets. The chances of success are reduced to 40 per cent, but it is fervently hoped that at least some data will be sent back, because if Voyager 2 fails there will be little hope of learning much more before the end of the century. No more probes to the outposts of the Solar System are planned, and by the time they can be made ready we will have lost the chance of using the planetary alignment that was so useful to the Voyagers. Of course it will still be possible to send a spacecraft to Neptune, but it will take longer, because we will not have Jupiter, Saturn and Uranus to help it on its way.

Meanwhile, what do we actually know about Neptune?

It is visible with binoculars as a starlike point, and powerful telescopes show a small disk, blue rather than green as with Uranus. The

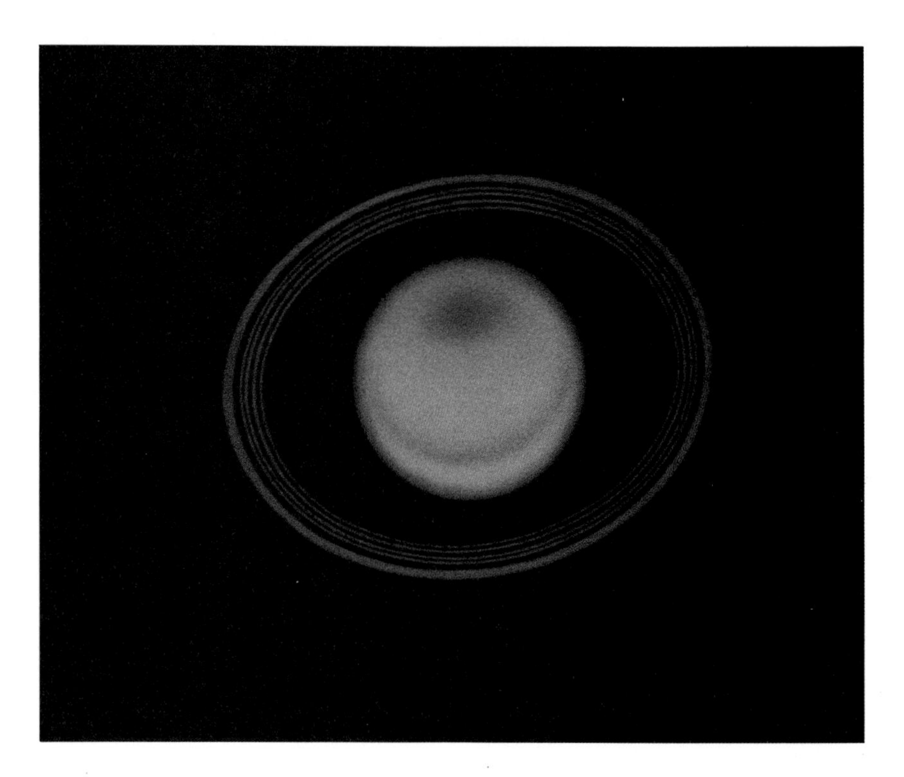

Impression of the rings of Uranus; since the rings lie in the plane of the planet's equator, the pole appears in the middle of the disk in this painting by Paul Doherty.

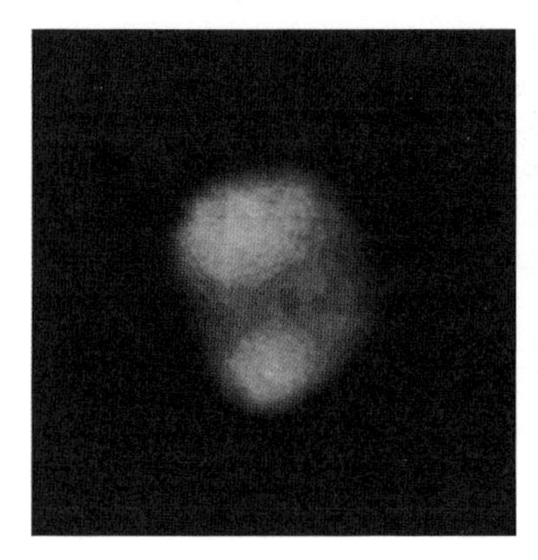
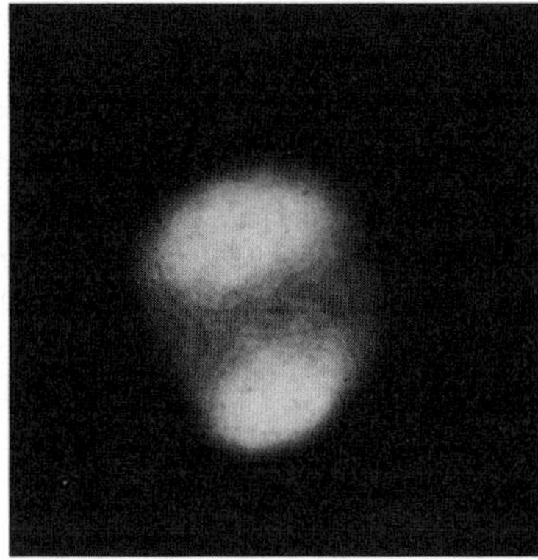
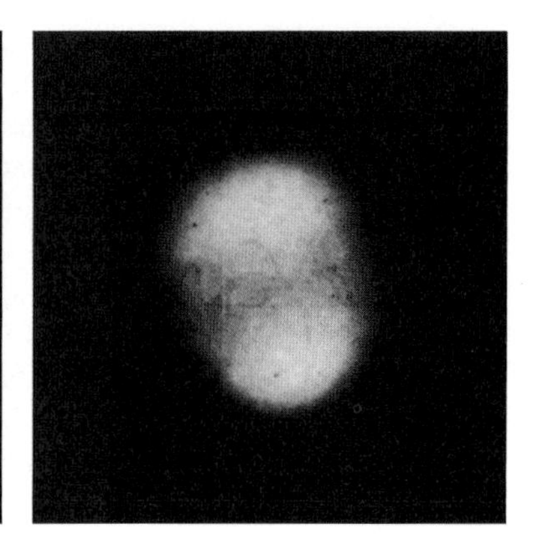

Neptune. These photographs were taken with a CCD (Charge-Couple Device) on the 61-inch telescope of the Catalina Observatory, U.S.A., by H. J. Reitsema, B. A. Smith and S. M. Larson of the University of Arizona on 5 May 1979. The pictures were recorded through a narrow filter which isolated the infra-red methane absorption band. High clouds of ice crystals in the northern and southern hemispheres produce the two bright areas, while the absence of any haze layer in the equatorial region reveals a deeper layer of methane gas. During the period covered by the three images, the bright features moved towards the eastern limb of the planet. Most photographs of Neptune are featureless; these are the best ever produced, and are unlikely to be surpassed until the Space Telescope is in orbit.

orbital period is $164\frac{3}{4}$ years; its mean distance from the Sun is 2793 million miles, and its rotation period is rather less than 24 hours, though, as with Uranus, we have no precise value. Neptune is somewhat smaller than Uranus, with a diameter of around 30,800 miles, but it is more massive, 'weighing' 17 times as much as the Earth.

In some ways Uranus and Neptune seem like twins, and there are indeed many points of resemblance, but they are not identical. The main difference is that Neptune has an internal heat-source, while Uranus has not (at least, none has been detected; there is always the chance that the heat is somehow or other blanketed inside the globe). This has been known for some time. At the high-altitude observatory on Mauna Kea in Hawaii, about which I will have more to say later, Dale Cruikshank measured the infra-red temperatures of the two outer giants, and found Neptune to be the warmer of the two, even though it is much further away from the Sun. There is also much more haze in Neptune's outer atmosphere, together with particles, which may or may not be ice crystals. Neither does Neptune share the peculiar calendar of Uranus; the axial inclination is 29° to the perpendicular, which is less than 6° more than our own.

Neptune has been known ever since 1846. It was actually 'discovered' before it was 'seen', and I must delve back temporarily into history, because it is important in what follows.

When Uranus had been identified, in 1781, mathematicians set to work to calculate its orbit. Unfortunately, Uranus refused to behave; it persistently strayed from its predicted path. Something unknown was pulling it out of place, and this 'something' could only be a planet, moving at a greater distance from the Sun. John Couch Adams in England and Urbain Le Verrier in France, working independently and unknown to each other, undertook some cosmic detective work, and calculated the likely position of the perturbing planet. They were right; when telescopes were aimed at the indicated position, there was Neptune.*

Shortly afterwards a skilled English amateur, William Lassell, discovered that Neptune has a large satellite, now named Triton. And over a century later Gerard Kuiper detected a much smaller attendant, Nereid.

*It has since been shown that Neptune was seen in 1612 and 1613 by no less a person than Galileo, the first great telescopic observer. At that time Neptune was close to Jupiter in the sky, and Galileo recorded it twice, though naturally he took it to be a star.

A third satellite was discovered in 1981, by Taylor's occultation technique, but its orbit is not yet well known.

Triton is an enigma. It is larger than any of the satellites of Uranus, and certainly larger than the Moon; its diameter may be over 3000 miles, in which case it is the equal of Mercury. Its mean distance from Neptune is 220,000 miles, but – and this is the significant point – it moves round the planet in a wrong-way or retrograde direction, taking 5 days 21 hours to complete a full circuit. All the other retrograde satellites in the Solar System are small, and probably asteroidal, so that Triton is unique. The surface temperature has been given as −360 °F., so that the rock-ice surface may be coated with a layer of methane frost. The other confirmed satellite, Nereid, is much smaller, and has a very elliptical orbit so that its distance from Neptune varies over a wide range.

I stress the retrograde motion of Triton because it seems to me to be important. First, what effects will it have on ring formation? Jupiter, Saturn and Uranus all have ring systems, admittedly of different types, and it has been tacitly assumed that Neptune also has a ring. I am not so sure. The presence of a massive satellite orbiting in a direction opposite to that in which Neptune spins will make for unstable conditions. Calculations made in 1981 indicate that a ring cannot be ruled out, but I am prepared to go on record as suggesting that Neptune, alone of the giants, will turn out to be ringless. We may find out in August 1989, provided that Voyager 2 is still operating. If I am right, I will draw attention to my prediction; if not, I will quietly forget that I ever made it!

But why does Triton move in this way, and what can account for the eccentric orbit of Nereid? It has been suggested that, in the distant past, some tremendous upheaval took place in that part of the Solar System, and this brings me on to the story of Pluto, where again there have been some interesting developments during the past few years.

The story really goes back to the time of Percival Lowell, more than 80 years ago. I have already commented that it is unfair to remember Lowell only for his admittedly rather wild theories about Martian canals; he was a good mathematician as well as an excellent organizer, and he became interested in the problems of the outer planets. Neptune had been tracked down because of its effects upon Uranus, but things were still 'not quite right', and Lowell came to the conclusion that yet another planet might exist. He calculated a probable position, and then started to search.

Of course he had the Flagstaff telescopes at his disposal, and he could also make use of photography. The method was to photograph the same star-field twice, with an interval of a few nights. The stars would remain in the same relative positions, but a moving planet would not; it would shift, and so betray its true nature. The search was difficult, partly because the planet was obviously going to be faint, partly because its position was extremely uncertain, and partly because of asteroids, which move in orbits closer to the Sun and complicate the issue. Lowell did his best, but when he died, in 1916, the planet was still unfound. Another hunt, organized in 1919 on the basis of calculations by the American astronomer W. H. Pickering, met with an equal lack of success.

There the matter rested for some time, but it was always in the mind of the Flagstaff team, and in 1928 it was decided to make another effort. The 24-inch refractor was not really suitable, because it had too small a field of view, and so a new telescope was ordered specially for the hunt. It was a refractor with a 13-inch object-glass, and its wide field meant that a comparatively large area of the sky could be photographed with a single exposure. It was paid for by Percival Lowell's brother, A. Lawrence Lowell, and is still known officially as the Lawrence Lowell Telescope. By the spring of 1929 it was installed at Flagstaff, and the work began.

The 'hunter' was Clyde Tombaugh, a young amateur astronomer who had been brought up on a Kansas farm and had no qualifications whatsoever apart from an exceptional amount of enthusiasm, patience and skill. Systematically he began to search, photographing on every clear, dark night and then comparing his plates by putting them into a device known as a blink-comparator. Here, the two plates are viewed in quick succession by switching the optical system. The stars will remain motionless, but a moving object will jump, and will appear as a blinking spot (hence the name). The procedure was remarkably tedious, and in the end Clyde Tombaugh blinked more than 90,000,000 stars.

On 18 February 1930, while blinking plates which had been exposed during the previous month, he came across a dim speck which moved by exactly the right amount. In his own words: 'A terrific thrill came over me. I switched the shutter back and forth, studying the images. Oh! I had better look at my watch and note the time. This would be an historic discovery. Estimating my delay at about three minutes, it would place the moment of discovery very close to four o'clock. For the next 45 minutes or so I was in the most excited state of mind in my life.'

All the checks were positive. Tombaugh walked over to the office of the Observatory Director, Vesto Slipher, and broke the news. Over the next clear nights the object was identified telescopically, and again photographed; when there was no longer any doubt, an announcement was made. The date was 13 March, 149 years after Herschel's discovery of Uranus and the 75th anniversary of Lowell's birth. After some discussion the name chosen was that of Pluto, God of the Underworld. It

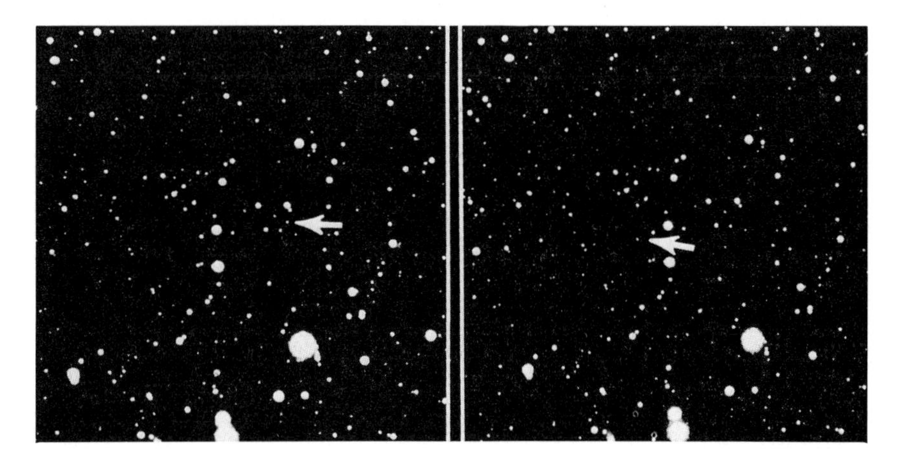

Pluto; 21 and 29 January 1930; photographs by Clyde Tombaugh using the 13-inch refractor at the Lowell Observatory, Flagstaff. Pluto is arrowed.

The 13-inch Lawrence Lowell telescope used by Clyde Tombaugh in the hunt for Pluto. When I took this photograph, in 1981, the telescope looked virtually the same as it did in 1930, but now that it has been moved from Flagstaff to Anderson Mesa it is correspondingly more effective.

OPPOSITE: Clyde Tombaugh: March 1980, 50 years after the announcement of the discovery of Pluto. I took this photograph in his garden at Las Cruces, New Mexico; part of the mounting of his 17-inch reflector is visible in the background.

was certainly appropriate: Pluto was a gloomy god, and the planet named in his honour must be a gloomy place.

The 13-inch telescope is still in full operation, though it has now been moved from the main headquarters at Flagstaff and, together with the other large telescopes, has been installed at Anderson Mesa, some miles away. Conditions there are better than at the old Observatory, because Anderson is higher and further away from the lights of the growing town. Last time I went there, in 1980, in the company of Clyde Tombaugh, the 13-inch looked exactly the same as it must have done in 1930, and it has certainly earned a place in scientific history.

Pluto was soon found to be an exceptional world. Lowell had failed to identify it because it was unexpectedly faint (in fact, it was subsequently tracked on two plates taken with the 24-inch as early as 1908). It was visible only as a dot of light, so that it could not be a giant; it seemed as though it must be smaller than the Earth, and moreover it had a strange orbit, bringing it within the path of Neptune for some time during each 248-year period, though there is no danger of a collision with Neptune because Pluto's orbit is tilted at the unusually high angle of 17°. It next reaches perihelion in 1989, and between now and 1999 it temporarily forfeits its title of 'the outermost planet'.

Within a few months after the discovery, doubts began to creep in. If Pluto were so small, and presumably so low in mass, it could hardly cause any measurable irregularities in the movements of giants such as Uranus and Neptune – and yet it had turned up very close to the position

which Lowell had worked out. There seemed to be three possibilities, none of them very convincing. Pluto might be exceptionally dense, more so than solid iron. It might be much larger than the diameter measures indicated, with a shiny surface which reflected the sunlight over a limited region only. Or Lowell's correct prediction might have been sheer luck.

The only way to find out was to make an accurate measurement of Pluto's size, which was a difficult matter because not even the Palomar reflector would show a proper disk. The only real possibility was to use Gordon Taylor's occultation technique, but Pluto looks so tiny, and is so sluggish in movement, that occultations by it are extremely rare; neither are they predictable within the required accuracy, since an error of less than a second of arc, either in Pluto's position or that of the star, would make all the difference.

The need, then, was to check photographs of Pluto taken over long periods, so that the predictions could be improved. Dr James Christy, at the United States Naval Observatory – also at Flagstaff, though not connected with Lowell – began to check the plates that had been taken with the large reflector there from 1965 onwards. He also took photographs on his own account, and in June 1977 he noticed something very curious indeed. Pluto's image was not circular; it appeared to be elongated, or even double. Christy made even closer studies of the plates, and decided that there could be only one answer. Pluto was not a single body, but was accompanied by a relatively large satellite, so that in the blurred pictures available it looked like a dumbbell with one bell larger than the other.

Christy contacted Dr John Graham, who was then using the 158-inch reflector at the Cerro Tololo Observatory in Chile, and asked him to check. The double appearance was seen again, and further confirmation came from the high-altitude Mauna Kea Observatory in July 1978. Yet for some time the discovery was greeted with considerable scepticism; it was suggested that the effects were spurious, or even that Pluto was genuinely elongated in shape. Final proof of the existence of the satellite – by that time named Charon, after the ferryman who took departed souls across the River Styx into the Underworld – was obtained on 20 June 1980, when D. Bonneau and R. Foy, using the Mauna Kea 144-inch reflector, obtained a photograph which showed the two bodies clearly separated.

Charon is not a normal satellite. Its diameter is more than a third that of Pluto itself; perhaps 900 to 1000 miles, as against 2000 to 2500 for Pluto. Also, it moves at less than 13,000 miles from Pluto's surface, and the revolution period of 6 days 9 hours 17 minutes is the same as the axial rotation period of Pluto, so that it is, so to speak, 'locked'; to a Plutonian observer it would stay motionless in the sky (from Pluto's far side, of course, Charon would never be seen at all). We are dealing not with a conventional planet-and-satellite arrangement, but with a double or binary planet.

For that matter, can Pluto be ranked as a true planet at all? The answer today seems to be, 'No', because it is too small, perhaps smaller than the Moon and certainly smaller than Triton. It may consist chiefly of ice,

The first photograph to show Pluto and Charon individually; it was obtained using a complicated technique known as 'speckle interferometry'. This involves taking a large number of short-exposure photographs, enhancing the images electronically, and then 'averaging them out', either by a computer or by a special optical system. The method is very powerful, and of course conditions on top of Mauna Kea are excellent. Ample confirmation of the separate existence of Charon has since been obtained.

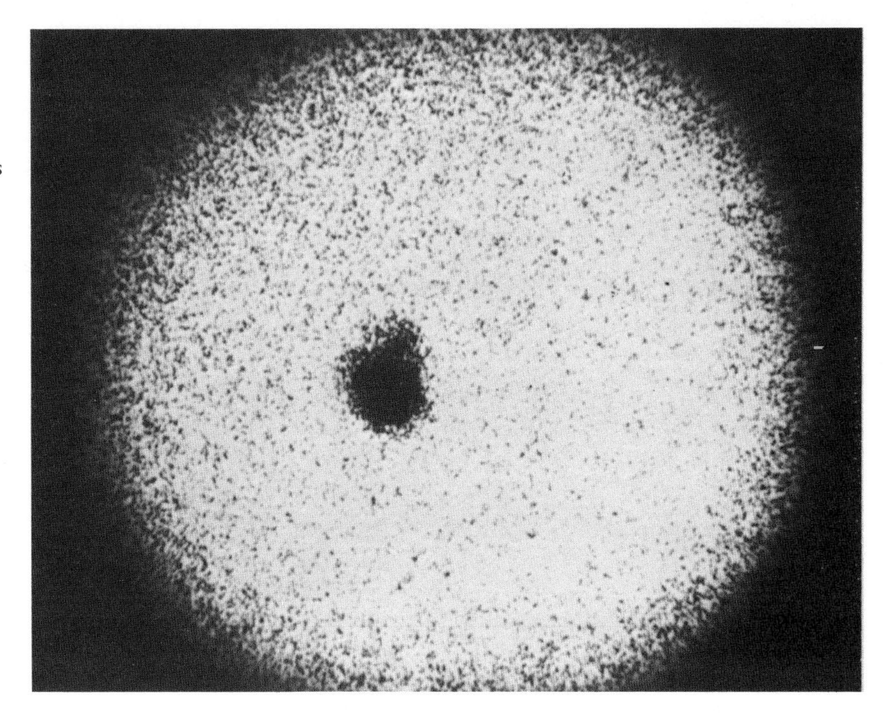

with a surface coating of methane frost, and a very tenuous methane atmosphere.

Over 20 years ago the Cambridge astronomer R. A. Lyttleton revived an old idea that Pluto used to be a satellite of Neptune, and was wrenched free by some invading body so that it moved off in a path of its own, while Triton was thrown into a retrograde orbit round Neptune and Nereid into a very eccentric one. The discovery of Charon seems to rule this out, because it is hard to visualize a satellite of a satellite, and if Pluto and Charon had stayed close to Neptune for long they would have been separated. On the other hand it is quite likely that Pluto is only one of a number of remote, asteroidal-type objects, and this was supported by another discovery made in 1977, this time by Dr Charles Kowal at Palomar.

Using the 48-inch Schmidt telescope there (not the 200-inch), Kowal was carrying out a systematic survey of the Solar System the main aim being to identify distant comets. The 48-inch was particularly suitable for this work, because it has a wide field of view; it collects its light by using a spherical mirror, and it has a special correcting lens at the top of its tube, so that a Schmidt system uses both a lens and a mirror. (The principle was devised in 1930 by the Estonian optician Bernhard Scnmidt, who had originally been experimenting with explosives, but blew off one of his arms and then turned to astronomy as being rather safer.) In October 1977 Kowal took two plates on which he identified a moving dot that did not seem to be either a comet, an asteroid or a trans-Neptunian planet. It was soon found to be moving in an orbit lying mainly between those of Saturn and Uranus, and to be perhaps as much as 400 miles in diameter; very recent work (1982) indicates that it is probably rocky

Pluto, photographed in 1930 by Clyde Tombaugh with the 13-inch refractor at Flagstaff. The photograph was taken on 2 March. The brilliant very over-exposed star is Delta Geminorum, sometimes known by its old proper name of Wasat.

rather than icy. It was named Chiron, after the wise Centaur who taught Jason and the other Argonauts of mythology (perhaps a rather unfortunate choice, because it is so like Charon, the satellite of Pluto).

What is unusual about Chiron is its position in the Solar System. Only when near perihelion does it come within Saturn's orbit, and this region, between the paths of Saturn and Uranus, is the last place where one would have expected to find an asteroid, if indeed Chiron is really an asteroid of the usual type. It takes 50 years to go once round the Sun, and it is naturally very faint. It may be only the brightest member of a whole swarm of trans-Saturnian bodies, and this swarm could even include Pluto. In any case we must, I am afraid, dethrone Pluto from its planetary status. It is simply not large enough or massive enough.

We return to the original problem: Why did Pluto fit in so well with Lowell's prediction, and for that matter, one of Pickering's? The 'good luck' theory is popular, but I do not believe it for a moment, because it would be too much of a coincidence. More probably there is another planet beyond Neptune, responsible for the effects on which Lowell based his calculations; but even if it exists, we have no real idea of where it is, and it will be so faint that its discovery will be difficult. To undertake a systematic search would mean monopolizing the observing time of a large telescope over a very long period, with no guarantee of success in the end and no reasonable astronomer is likely to do that. So for the moment, the mystery remains.

Even if Pluto is not a true planet, nothing can detract from the extent of Clyde Tombaugh's achievement. He is still very active in research, I am glad to say, and is Professor Emeritus at the University of Las Cruces in New Mexico. There, in March 1980 – exactly half a century after Pluto's discovery – a conference was held in his honour. Clyde Tombaugh was awarded the Regent's Medal, the highest honour that the University can bestow, and an asteroid (No. 1604) was named after him, leading him to comment that at least he now had a piece of real estate that nobody could touch! It was a great occasion, and I am very glad to have been there.

It is a pity that neither of the Voyagers will pass near Pluto, and it is now unlikely that any spacecraft will encounter it for a long time ahead. But Pluto will wait; and sooner or later we will be able to find out something more positive about this curious, dimly-lit little world at the edge of the main Solar System.

# 9 Space Nomads

If you happen to be visiting Flagstaff, either to go to the Lowell Observatory or for any other reason, you will be within easy driving reach of what was described by the great Swedish scientist Svante Arrhenius as 'the most interesting place in the world'. It is a vast hole in the ground, over 4000 feet in diameter and 570 feet deep. Its origin is no mystery; it was blasted out of the desert by a giant meteorite, which landed there some 22,000 years ago.

As you approach it, leaving the main highway and driving down Meteor Crater Road, you will see nothing impressive. The walls of the crater rise to only a very modest height above the desert, and could easily be mistaken for a mere ridge. But as soon as you stand on the rim, the view is breathtaking. The walls are steep; there is a 'trail' down to the floor, but it involves a good deal of scrambling, and the return climb under a broiling Arizona sun can be an exhausting business, as I have good reason to know (last time I went there I was taken down in a helicopter, together with our television equipment, which was a great deal easier; for some time I was absolutely alone inside the crater, and felt very lonely and cut-off!). The remains of the meteorite itself are thought to be buried under the south wall, though the missile broke up as it plunged through the lower atmosphere, and deposited many fragments – some of which are on display in the museum that has been built on the crater rim.

Meteor Crater (it really should be Meteor*ite* Crater) was first seen by white men in 1871, and was originally assumed to be volcanic, but from 1903 onwards a prospector named Daniel Barringer made a careful study of it, and established that it had been caused by something from outer space. Many meteorites are made of iron; iron is valuable, and Barringer did his best to locate the meteoritic mass, so that he could mine it. Fortunately for posterity he failed, and the site has now been taken over as a National Monument, so that the crater is protected and there is no danger of any future mining operations there. It has even been nicknamed 'America's oldest museum'.

Nobody can have seen the fall, because so far as we know there were no men in North America 22,000 years ago, but it must have been cataclysmic. Striking the desert at about 30,000 m.p.h., the missile could produce an explosion as powerful as the setting-off of half a million tons of T.N.T. Any life in the area would have been wiped out, both by the blast and by the searing heat, while trees would have been blown flat.

*Top:* Meteor Crater, Arizona; a photograph which I took from an aircraft in 1964. *Lower:* Looking across Meteor Crater. I took this picture when I was standing on the rim just in front of the museum. The meteorite itself is probably buried deep under the wall at the far side of the crater.

Meteor Crater is not unique. Other impact craters are scattered here and there; for instance there is one at Wolf Creek in Australia, smaller than the Arizona Crater and less imposing, but unquestionably of external origin. Yet it is only too easy to jump to conclusions, and many of the so-called impact craters in official lists are more probably of internal origin. Such is the Vredefort Ring in South Africa, some way north of Pretoria. It is 30 miles across, and two villages (Parys, and Vredefort itself) lie inside it; even when you survey it from a helicopter, as I did a few years ago, it is not really well-marked, but the crater form is easy to recognize, and it is listed in many catalogues of alleged impact structures. In fact it is certainly volcanic, as has been established by long-continued studies of it carried out by L. O. Nicolaysen, Duncan Stepto and other South African geologists. The same may be true of the large craters on the Canadian Shield, though here the evidence is much less clear-cut.

However, there is no doubt that meteorite craters exist on the Earth, as well as on the Moon and other bodies. What, then, are the dangers of another major fall?

Two large meteorites have come down during the present century, one in 1908 and the other in 1947, both in Siberia. The first was seen to fall, and during its descent it outshone the Sun. The area, the Tunguska region in the far north, was mercifully uninhabited, but when the first expeditions reached the site nearly 20 years later they found that pine-trees had been flattened over a wide radius, so that if the missile had

The world's largest known meteorite, at Hoba West, near Grootfontein in Southern Africa. Photograph by Ludolf Meyer

landed on a city the death-toll would have been enormous. The 1947 meteorite broke up during its descent, and here too the landing area, in the general region of Vladivostok, was uninhabited. There is in fact no well-authenticated case of anyone having been killed by a meteorite. (The only casualty has been a dog, which was in the wrong place at the wrong moment when a small meteorite came down in Egypt many years ago.)

Recent investigations seem to show that the 1908 object may have been the nucleus of a small comet rather than an ordinary meteorite, but the whole subject of cosmical collisions has been widely discussed lately, and has even been linked with an event which took place 65,000,000 years ago: the sudden extinction of the great dinosaurs which had been lords of the Earth for so long.

Most people are familiar with dinosaurs. If you want to make their acquaintance, you need do no more than go to any major natural history museum, where you will see their skeletons as well as skilful reconstructions. Contrary to popular belief, most of them were harmless and vegetarian, though there were some notable exceptions – such as *Tyrannosaurus*, arguably the most ferocious beast ever to have lived. For many millions of years the dominance of the dinosaurs was unchallenged, but at the end of the Cretaceous Period they suddenly disappeared. This is lucky for us, because otherwise mammals would have had little chance to develop; but what killed the dinosaurs so abruptly?

All sorts of theories have been proposed, some of them more plausible than others. One favourite explanation is the explosion of a supernova, a stellar outburst of unbelievable violence that soaked the Earth with lethal radiation and wiped out all but the hardiest forms of life. Today, however, probably the best-supported theory is that a body, perhaps half a dozen miles in diameter, impacted the globe, plunging deep into the crust and producing large-scale disturbances in the atmosphere. Such an event would not vaporize the Earth, but it might well usher in a prolonged period of darkness and constant cloud.

What effect would this have? Clouds blanket in heat, and if the world suddenly warmed up the huge but unintelligent dinosaurs would be unable to cope with the changing conditions. There are various pieces of supporting evidence, and in particular there seem to be unusually large amounts of the element iridium in the rocks laid down at the very end of the Cretaceous Period. Meteorites can sometimes bring iridium with them.

The great British astronomer, Sir Fred Hoyle, has come out in support of the theory that the dinosaurs were wiped out because of a meteoritic impact, but he believes that the result was a cold spell rather than a warm one, because the material sent up into the high atmosphere would absorb the Sun's heat. The luckless dinosaurs would not be able to see in the prevailing gloom; they would be affected by flooding and a shortage of food, so that they could not survive, whereas the small mammals which at that time co-existed with the dinosaurs were much more versatile. No meteorite smaller than about six miles in diameter could produce so drastic an effect, which is why the impact of 65,000,000 years ago was so exceptional.

OVERLEAF: Impression of the double asteroid Herculina. The main object is 150 miles in diameter, and the 'satellite' 30 miles; the distance between the two is about 600 miles. The Sun is shown to the lower left. Painting by Paul Doherty

The minor planet Eros (*arrowed*), photographed by Paul Doherty during its last close approach, that of 1975. The two bright stars are the 'Twins', Castor and Pollux in Gemini.

Could it happen again? Yes, says Hoyle; eventually it must. We could all be plunged into a new Ice Age without any prior warning, and the effects on civilization would be dire. His remedy is to take immediate precautions by building powerful machines to stir up the waters of the oceans, mixing them and so preventing very cold water from remaining at a very low temperature just above the ocean bed. If we had a warmer sea, the Earth's temperature after a Cretaceous-type strike would be maintained at a reasonable level for some time – long enough, Hoyle believes, for the climate to revert to something like normal.

Whether this is regarded as necessary or not, it is certain that collisions will happen occasionally, and new supporting evidence has come from observations of the type known as Apollo asteroids. A few of them have been known for some time, but it was not until 1973 that a systematic hunt for them was started, due to Eugene Shoemaker and Mrs Eleanor Helin in California.

Most of the asteroids are orderly and well-behaved, keeping strictly to the wide gap in the Solar System between the orbits of Mars and Jupiter. Only one (Ceres) is over 500 miles in diameter, and only one (Vesta) is ever visible with the naked eye; most of the rest are extremely small, and many may be less than a mile across. There is probably no difference between a large meteorite and a small asteroid; it is purely a question of terminology.

Some of the asteroids are moving in orbits well outside the main zone. The Trojans, for example, move in the same orbit as Jupiter, keeping either well ahead or well behind. Another asteroid, Hidalgo, has a very eccentric orbit that takes it out nearly as far as Saturn. We must also remember Chiron, though whether or not this peculiar little world is truly asteroidal remains to be seen.

Of more immediate interest are those asteroids that swing inside the path of Mars. The first of them, Eros, was found in 1898. It is shaped rather like a sausage, 18 miles long by 9 miles wide, and when it makes its closest approaches to the Earth, as last happened in 1975, it may come to within 15,000,000 miles of us. But it was Apollo, discovered by Karl Reinmuth on a photographic plate exposed in 1932, which made history, because its orbit crossed not only that of Mars, but also that of the Earth. Amor, found later in the same year, was of the same order of size, although its orbit did not actually intersect the Earth's. Next, in 1936, came Adonis, no more than a mile across, and then Hermes, a tiny object which brushed past us in 1937 at a mere 485,000 miles, which is less than twice the distance of the Moon. When the news was released to the Press, it made quite a stir. One London paper made it the lead story, under the headline WORLD DISASTER MISSED BY FOUR HOURS AS TINY PLANET HURTLES PAST!

It is not easy to keep track of Apollo-type asteroids, because they are subject to being 'pulled about' by the gravitational forces of larger bodies. Hermes has never been seen again, and we have no idea where it is now; Apollo was temporarily mislaid, but turned up once more in 1973, while Adonis was finally recovered in 1977 after a prolonged search. In 1981 the Arecibo radio telescope was used to make radar contact with two members of the group, Quetzalcoatl and Apollo itself.

Another maverick is Icarus, whose path carries it within 17,000,000 miles of the Sun, closer than Mercury; it must then be red-hot, though at aphelion it moves out well beyond Mars. Asteroid No. 1685, Toro, has an orbit not very different from that of the Earth, and makes regular close approaches, though it can never come nearer than a million miles, and is certainly not a 'second moon', as was suggested some time ago; it revolves round the Sun, not round the Earth. And there are three asteroids, Aten, Ra-Shalom and another yet to be named, whose orbits lie wholly inside that of Mars.

Despite the small size of these Apollo-type objects, they could be a potential danger. If one of them hit the Earth it could produce a crater at least a dozen miles across, and since the energy released would be tremendous the devastated area would be wide. An impact in, say, Manchester would leave little of Glasgow in the north or Bristol in the south, and the force would be at least four times more violent than that of the Krakatoa volcanic eruption in 1883. If the impact were in the ocean, waves in the hundred-foot category would cause worldwide damage.

If there were a threat of an asteroid collision, could we do anything about it? There might be adequate warning, and a study carried out in the United States has indicated that it would be possible either to divert the unwelcome visitor or else shatter it by means of a well-aimed nuclear bomb. In any case, we are entitled to hope that we would deal with the situation better than the dinosaurs did.

I have said that the 1908 Siberian missile may have been the central part or nucleus of a comet. What then can we say about comets in general, and do they also pose threats of any sort?

Comets are fragile, insubstantial bodies compared with planets, and almost all the cometary mass is concentrated in the nucleus, which may be several miles in diameter; our best available evidence for this was produced in 1981, when Paul Kamoun and his team at Arecibo managed to bounce radar pulses off the nucleus of a famous periodical comet, Encke's, and found the diameter of the relatively solid centre to be just under two miles, though admittedly with some degree of uncertainty. As comets move round the Sun they leave dusty trails in the form of meteors, and it is these which cause the familiar shooting-star streaks in the sky.

Most comets move in very elliptical orbits, and since they are not visible except when in the inner part of the Solar System we cannot follow them constantly. There are many comets whose periods are short – only 3.3 years in the case of Encke's – and with a single exception, all are too faint to be visible with the naked eye. The really brilliant comets that have been seen throughout history have much longer periods, and return to the neighbourhood of the Sun only at intervals of many centuries, so that we cannot predict them, and never know when or where to expect them. Several were seen during the 19th century, and the comets of 1811 and 1843 were brilliant enough to be visible in broad daylight, but there has been a disappointing dearth of them in recent years. A few have been bright enough to cause general interest –

Five photographs of the bright comet of April 1957, Arend-Roland, taken with the 48-inch, Schmidt telescope at Palomar; 26, 27, 29, 30 April and 1 May. In the first three pictures the famous 'spike' is shown. This was not a genuine 'front tail', but was due to material scattered in the comet's orbit being suitably illuminated by the Sun. Arend-Roland will presumably return one day, but not for many thousands of years.

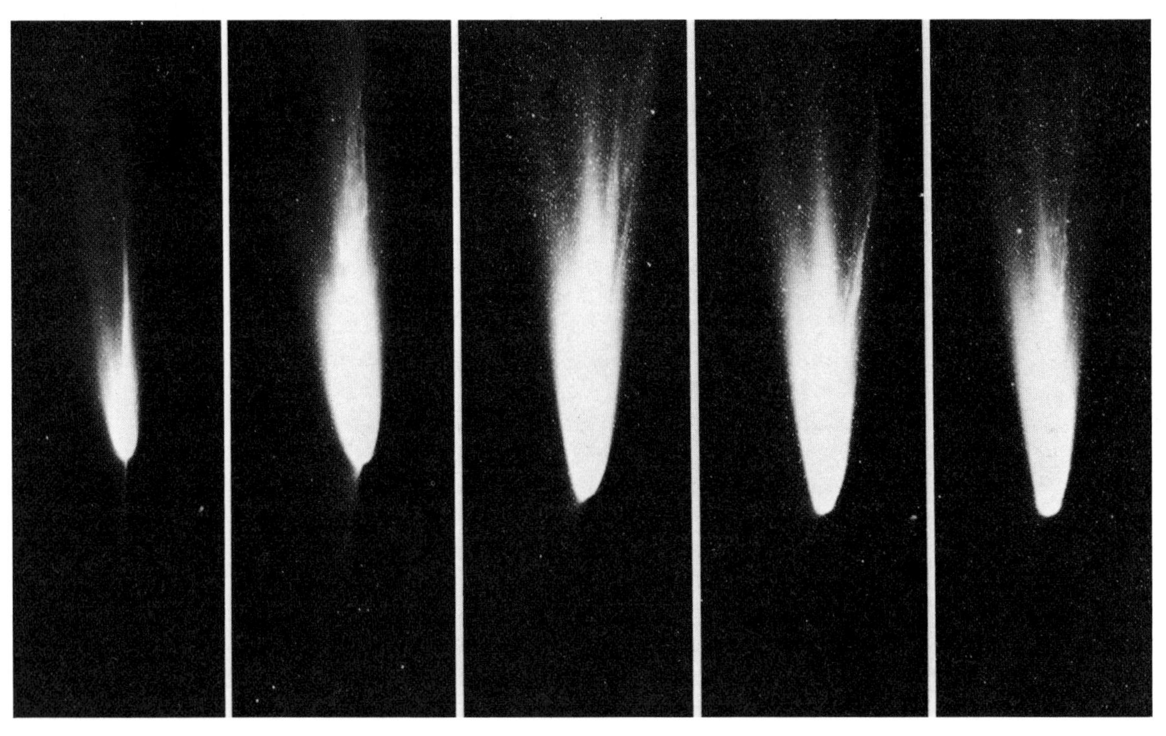

Kohoutek's Comet, as seen from the last crew of Skylab: Astronauts Gerald Carr, Edward Gibson and William Pogue.

the Arend-Roland Comet of April 1957, for example, was one which I remember vividly, because it was the subject of my very first 'Sky at Night' television programme.

The most disappointing comet of modern times was that of 1973. It was discovered by Dr Lubos Kohoutek, a Czech astronomer working at the Hamburg Observatory in Germany, and since it was then a long way away it was expected to become really brilliant as it neared the Sun. Preliminary estimates suggested that at its peak it might rival the Moon in brightness, in which case it would be seen during the day. At the time, weird pamphlets were issued by religious extremists – I still have one entitled *The Christmas Monster*, in which it was claimed that Kohoutek's Comet would either destroy the world or at least leave it in very poor shape. In the event, the comet failed miserably as a spectacle. It was only just visible with the naked eye, and it was not nearly so conspicuous as another comet, Bennett's, seen three years earlier. Since Kohoutek's Comet will not come back for something like 75,000 years, it will have no chance to make amends so far as we are concerned. However, important studies of it were carried out by the three astronauts then manning the American Skylab station, and the comet was found to be shrouded in an immense envelope of tenuous hydrogen gas.

One comet, recorded in 1979, seems to have met with a sad fate. It was detected on 30 August from a special instrument termed a coronagraph, designed to photograph the Sun's outer atmosphere, carried in a U.S. Defence Department satellite, P78–1. The comet was very close to the

Bennett's Comet of 1970, photographed by the discoverer – Jack Bennett – from Pretoria in South Africa. This was one of the most beautiful comets of modern times, and though not 'great' it was certainly conspicuous; as seen from Sussex its nucleus became as bright as the first magnitude – that is to say, equal to a star such as Altair. It has an immensely long period, so that it will not be seen again for many centuries.

Sun, and succeeding photographs taken over the next few hours showed that it was apparently heading towards the Sun's centre on a direct collision course. The tail reappeared after the passage behind the Sun, gradually diffusing, but the head did not, so that either the comet disintegrated because of the intense heat or else there was an actual collision. The comet had not been seen earlier because it had been unfavourably placed in the sky, and it was never seen again, so that we may have witnessed what can only be termed a cosmic suicide.

Comets used to cause general alarm, and were regarded as unlucky. There is no obvious reason for this, because a comet is so insubstantial, and several times during recent years the Earth has passed through cometary tails without coming to the slightest harm. But in 1978 an entirely new idea was put forward by Sir Fred Hoyle and his colleague Dr Chandra Wickramasinghe, which gives a fresh slant to the old terror of comets.

Hoyle and Wickramasinghe have gone back to a modified version of Svante Arrhenius' old theory that life did not originate on Earth, but was brought here from outer space. They point out, quite justifiably, that the appearance of life on a previously sterile planet involves a whole set of very unlikely coincidences, and they feel that the best conditions are to be found well away from any solid body. Organic molecules are plentiful in space; according to Hoyle it is these which produce 'life', and he goes on to suggest that comets are the best vehicles for carrying the life-bearing substances around. He even goes so far as to claim that all the water in our oceans was dumped here by comets in the remote past.

According to this view, comets are still capable of depositing bacteria in the upper atmosphere, and some of these may be harmful, so that comets can actually spread epidemics of various diseases ranging from influenza to smallpox. At the moment smallpox has been virtually stamped out, but if Hoyle and Wickramasinghe are right the virus could return in the future, by way of a comet.

It is fair to say that medical experts in general have no faith in ideas of this kind, and I admit to being highly sceptical, but it is quite true that we do not yet know nearly as much about comets as we would like to, and in particular we have very little information about their nuclei. However, we may learn more in the near future, because the best-known of all comets, Halley's, is now approaching us, and will be at its best in the early part of 1986.

Halley's is the only bright comet with a period short enough for it to be predicted. It takes its name from Edmond Halley, the second British Astronomer Royal, who saw it in 1582, found that its orbit was the same as those of comets previously recorded in 1531 and 1607, and forecast that it would return in 1758, which it duly did (it passed through perihelion in the following year). Its period is 76 years, though the pulls of the various planets cause this to vary by a year or two either way. Old documents show that it was probably first noticed in China in 467 B.C.; it was certainly observed again in 11 B.C., and since then it has been seen at every return, the last two being in 1835 and 1910. (Note that there were two bright comets in 1910; one was Halley's, but the other, visible a month or two earlier, was much more brilliant, though it will not be seen again yet awhile, for its period must be thousands of years). At perihelion, Halley's Comet is much closer to the Sun than we are, but at aphelion it goes out well beyond the orbit of Neptune, and because it moves at its slowest when furthest out it is a naked-eye object for only a few months every 76 years.

As with all bright comets, Halley's develops a long tail as it nears the Sun, produced by material evaporated from the head. But it is the nucleus which interests us most, and plans for sending a spacecraft to it during the coming return were laid well ahead of time.

The first announcement came from NASA. The idea was to use the Space Shuttle to launch the comet probe during the summer of 1985, so that it would rendezvous with Halley in early 1986 at a relative velocity of 14,000 m.p.h.; the probe would carry a heavy dust shield to protect

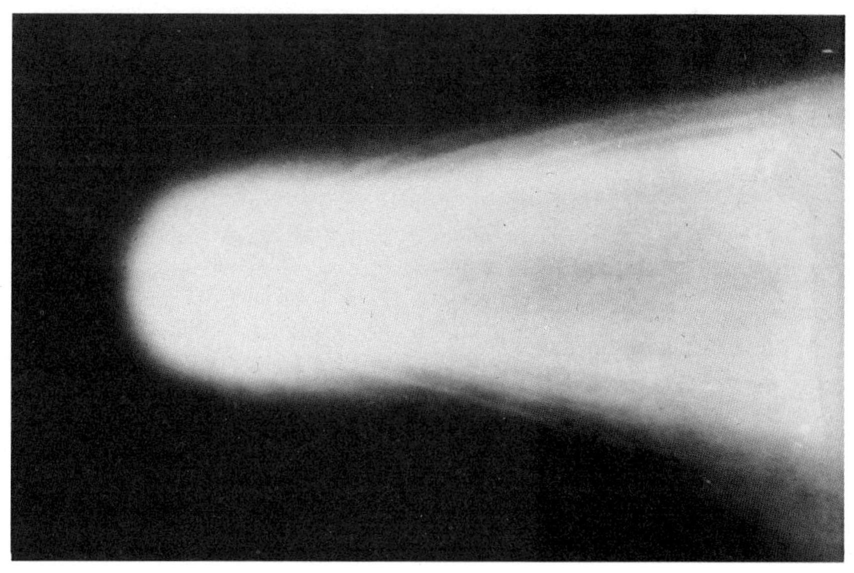

The head of Halley's Comet, photographed from Mount Wilson in 1910.

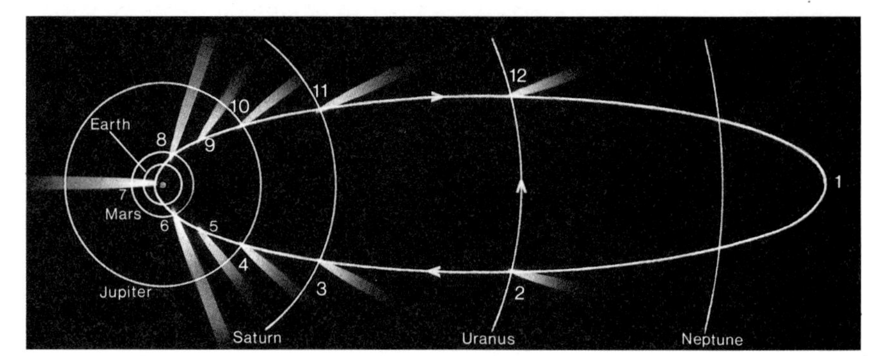


*Top:* Relative positions of the Earth and Halley's Comet during the perihelion passage of 1986.

**A** and **a** Comet and Earth at pre-perihelion closest approach, 27 November 1985.

**B** and **b** Comet and Earth at perihelion date 9 February 1986.

**C** and **c** Comet and Earth at post-perihelion closest approach, 11 April 1986.

**☊** and **☋** mark the positions of the ascending node and descending node respectively (i.e. the positions where the plane of the comet's orbit crosses the plane of the orbit of the Earth).

*Lower:* The orbit of Halley's Comet, with the orbits of the planets (excluding Pluto). The numbers indicate positions and dates as follows:

**1** 1948 and 2024
**2** 1977
**3** Mid-1983
**4** Late winter 1984
**5** Summer 1985
**6** Winter 1985
**7** 7 February 1986 and 19 April 1910
**8** Late spring 1986
**9** Late summer 1986
**10** Late autumn 1986
**11** Summer 1988
**12** 1999

(The references to seasons apply to the northern hemisphere of the Earth)

the cameras and other instruments designed to measure the chemical, atmospheric and particle properties of the comet, and all sorts of experiments were to be incorporated.

Two months before encounter, the cameras would begin their photographic surveys, and three hours before closest approach the probe would plunge into the cometary dust surrounding the nucleus, passing eventually within 1240 miles of the nucleus itself and sending back hundreds of high-resolution images, thereby treating us to our first good look at the heart of a comet. The encounter was scheduled to end in April 1986.

Sadly, financial problems intervened. NASA's funds were savagely cut back, and with great reluctance the comet mission was abandoned.

The decision caused profound gloom among astronomers. After all, Halley will not be back again until 2061–2, which is rather a long time to wait.

However, four probes remain. One, a fairly modest affair, comes from Japan, and is currently termed Planet A; it is scheduled for launching in 1985, and will be the first major Japanese effort in space, though they have launched several artificial satellites during the past few years. There is also Project Giotto, organized by the E.S.A. or European Space Agency, which should have a four-hour encounter, and is designed to approach within about 300 miles of Halley's nucleus. But the most sophisticated probes will be those of Project Vega, from Russia.

Two Vegas will be launched within a week or two of each other, with a third kept in readiness in case of any initial mishap. Both will pass by the dark side of Venus in March 1985, and will take the opportunity of sending instrumented packages into the atmosphere, bringing them down gently and obtaining pictures as well as miscellaneous data. (Another possibility is that of releasing small balloons as the probes plunge downwards, so that the balloons will float around in the sulphuric acid clouds and continue transmitting from different levels.) Meanwhile the Vegas themselves will hurtle on, propelled towards Halley by the gravitational force of Venus. On 8 March 1986 the first Vega should by-pass the comet, almost head-on, with the second following soon afterwards, racing past Halley's nucleus at a relative velocity of 50 miles per second. The minimum distance between the nucleus and the probes will be about 6000 miles, but the equipment carried – French and Hungarian, as well as Russian – will record details as little as 600 feet across.

Whether the Vegas will be able to carry out their full programmes remains to be seen. If they succeed, it will be a tremendous triumph, because so far the Russians have not attempted a two-target mission, and they have often had difficulties with long-range communication. We must wish them all success. Meanwhile, what will we on Earth see of Halley's Comet?

Unfortunately, conditions at this return are not good, and are much less favourable than they were in 1835 and 1910, because the Earth and the comet are in the wrong places at the wrong times. However, Halley should become visible with the naked eye in late November or early December 1985; it will pass through perihelion in February 1986, and will then reappear as it begins its long outward journey. For much of this period it will be in the southern hemisphere of the sky, fading as it comes northward again. According to prediction it will be at its brightest around 11 April, when its nucleus should equal Venus; it will then be at its closest to the Earth, between three and four million miles away. The estimated diameter of the nucleus is about three miles, and there should be a respectable tail, though this will shorten as the comet recedes from the Sun, and will be gone long before Halley has returned to the depths of the Solar System.

I have never seen a really spectacular comet – they have been disappointingly rare since 1910. Frankly, I am looking forward to seeing Halley, if only because so far as I and many others are concerned it will be the only chance.

# 10 The Inconstant Sun

This striking view of the Sun's corona was prepared from data supplied by the Solar Maximum Mission satellite, so named because it was operating during the time when the Sun was at the peak of its 11-year cycle of activity (mid-1980). The prominent line extending from the Sun towards the west is a coronal spike. Several other spikes can be seen to the south of the most prominent line. The spike extending from the densest part of the corona persists beyond 1,000,000 miles from the Sun's surface. Such spikes (or streamers) can extend out to nearly 10,000,000 miles. Shortly after this picture was taken, a solar flare occurred on the Sun; the portion of the corona above the flare was completely disrupted, and changed its shape in a matter of a few minutes. The white dots (which are camera imperfections) appear as roughly the same size as the Earth would do.

A few months ago I went to Kitt Peak, some way out of the bustling city of Tucson in Arizona. It is here that the United States authorities have set up what has become the accepted National Observatory. The Peak itself is around 7000 feet high; the drive up is pleasant, along a good if winding road; the scenery is magnificent, and the journey to the top does not take long.

Generally speaking the seeing conditions are good, which is why the site was selected from a list of 50 possible candidates. The preliminary negotiations began in 1950, and were unusual inasmuch as the Papago Indians had to be consulted. The area is within their reserve, and the sacred mountain Babuquivari, prominently seen from Kitt Peak itself, is regarded by them as the centre of the entire universe, while the gods live in nearby caves. The astronomers were very ready to co-operate; friendly relations were established, and the Observatory acquired the lease, after having made a firm promise that the sacred caves would never be disturbed.

The first dome on Kitt Peak was built in 1958. Others have followed, and by now there are seventeen in all, including those of the 24-, 36-, 84- and 90-inch reflectors. The most powerful telescope is the 158-inch reflector, one of the largest in the world and also one of the best. But to the visitor coming to the site for the first time, the scene is dominated by what looks like a huge, white inclined ramp. This too is a telescope, but a most unusual one. It has been constructed specifically to observe the Sun.

The problems of solar observation are different from those in all other branches of astronomy. Normally the main requirement is to collect every scrap of light available, but with the Sun there is plenty of light to spare, and the choice of design is much wider. Solar astronomers are not concerned merely with taking pictures of the Sun, with its bright clouds and darker patches. They are much more interested about spreading the light out into a spectrum, and the longer the focus of the telescope, the more spread-out or dispersed the spectrum can be.

Also, the equipment used for solar research tends to be heavy, and moving it around by fixing it to the end of a telescope is something to be avoided if at all possible. It is much better to have the sunlight coming in at a constant angle. Optical telescopes can manage this if they are suitably designed (according to a pattern known as the Coudé), but in the Kitt Peak solar telescope a rather different plan is followed.

At the top of the structure there is a mirror, 80 inches in diameter,

The solar telescope at Kitt Peak; a photograph which I took in January 1980.

known as the heliostat, which catches the Sun's light and directs the rays down a 600-foot tunnel in a fixed direction. The tunnel is slanted, and it is this which makes it look so like a ramp. At its bottom there is another mirror which is curved, instead of being flat like the heliostat. It is placed so as to reflect the rays back up the tunnel to a 'half-way' house, where yet another flat mirror sends the rays down through a hole in the tunnel into the solar laboratory below. This is where the analytical equipment is set up, and since the sunlight always comes straight down through the same hole there is no need to move anything in the laboratory. In fact, the only moving part of the whole optical system is the heliostat.

Most solar telescopes are smaller than this, but the large mirrors of Kitt Peak have marked advantages. The image of the Sun produced in the laboratory is almost three feet in diameter, and there is so much light that it can be spread out into a very long spectrum. With other solar telescopes the towers are usually vertical, but with the 600-foot Kitt Peak tunnel this would involve a truly dizzy height, which is why it has been set at an angle (even so, the heliostat is 100 feet above the ground). Temperature control is important inside the tunnel, because if it were not more or less constant, the air would become unsteady and ruin the telescope's performance. To ensure that this does not happen, 1700 gallons of cooling fluid are kept circulating between the inner and outer walls of the tunnel, so that the inside temperature stays constant to within a degree or two.

The heliostat at the top of the Kitt Peak solar telescope.

Bad weather can hit the Peak (I once went there to find myself deep in snow, with a howling gale screaming across the mountain), and it was originally planned to lower the heliostat into the top of the tunnel during storms. The cost of this extra facility was estimated at $75,000, but when this rose to $250,000, the Observatory authorities called 'Enough!' and the scheme was abandoned; instead, the heliostat is simply covered during bad spells, and it seems to be none the worse.

Going inside the telescope is most interesting. You climb up to the half-way house, to see one mirror at the bottom of the tunnel and the heliostat high above; when the Sun is not shining, there is very little light. A kind of miniature truck on rails is used to go up to the heliostat, which is certainly less exhausting than negotiating the dozens of steps.

Though the telescope was designed for solar work, it has proved to be so good that it can also be used for the stars, so that it is in action not only during the day but also for most of the night. Of course, its optics are not so large as those of some of the other telescopes – the main mirror is 'only' 60 inches in diameter – but once again the fact that the light is brought to focus at a fixed point is a great advantage. All in all, the telescope is probably the busiest on Kitt Peak. The gods living in the caves below Mount Babuquivari may well be proud of it.

The Sun is an ordinary star, and it is by no means distinguished in the Galaxy; astronomers rather unkindly relegate it to the status of a Yellow Dwarf. But to us it is all-important, and before going on I must pause to say more about it.

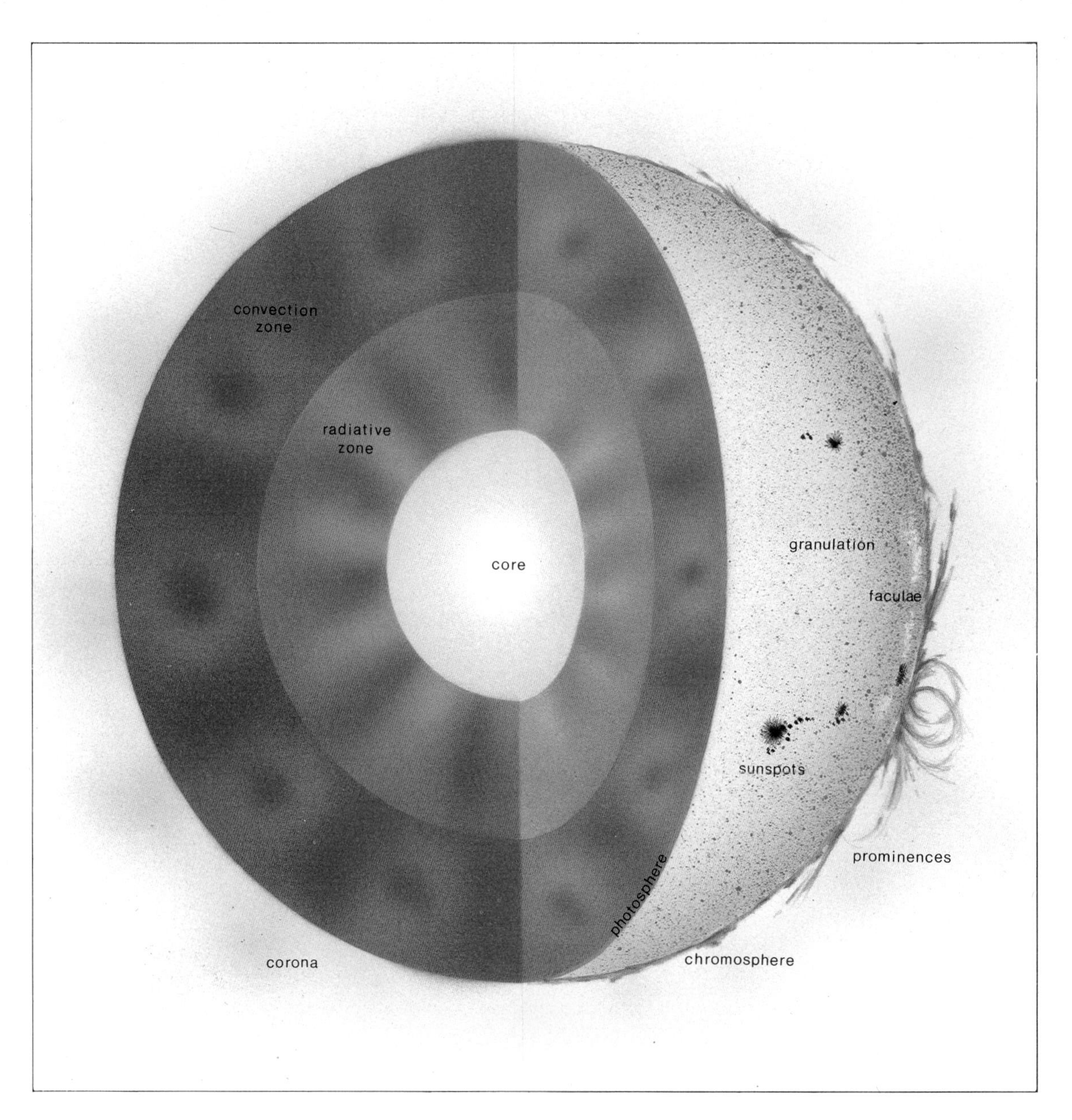


The diameter is 865,000 miles, and the globe is big enough to accommodate more than a million bodies the volume of the Earth. The surface temperature is almost 6000 °C.*, and the temperature at the core rises to the unbelievable value of around 15,500,000 °C. The Sun, of course, is gaseous throughout; the most plentiful element is hydrogen, with helium coming a rather poor second and the other elements very much less abundant, though by now more than 70 have been identified.

The Sun cannot be burning in the conventional sense of the term; it can easily be shown that a body the size of the Sun, made of coal and

*Stellar temperatures are always given in degrees Centigrade rather than Fahrenheit, and I propose to follow convention – particularly as the temperatures are too high to be properly appreciated, no matter which scale is used!

Cross-section of the Sun. The core, where the energy is being produced, is surrounded by a radiative zone, outside which comes a convective region. Sunspots are features of the bright surface or photosphere; so are the bright regions known as faculæ (from the Latin for 'torches') and the granular structure. Above the photosphere comes the chromosphere or 'colour-sphere', in which lie the prominences. Beyond the chromosphere comes the much more rarefied corona. It is notable that the outer boundary of the photosphere is comparatively sharp. Diagram by Paul Doherty

burning fiercely enough to give out as much energy as the Sun actually does, would turn itself into ashes in a few tens of thousands of years, and the Sun is much older than that. We know the age of the Earth to be between 4,500 million and 5,000 million years, and the Earth cannot have come into existence before the Sun.

Gravitational contraction, then? This is better, but still inadequate. The true solution was found in 1938, almost simultaneously by Hans Bethe in America and Carl von Weizsäcker in Germany. The essential clue is to be found in the behaviour of hydrogen.

At the Sun's core, the pressure is about 100 million times that of the Earth's air at sea-level, and, as we have seen, the temperature is fantastically high. Under these conditions, the nuclei or central parts of the hydrogen atoms are banding together to make up nuclei of the next lightest element, helium. It takes four hydrogen nuclei to build one nucleus of helium, and each time this happens a little energy is set free and a little mass is lost. It is this released energy that keeps the Sun shining, and the mass-loss amounts to 4,000,000 tons per second, though I hasten to assure you that there is no need for alarm. The total mass of the Sun is $2 \times 10^{27}$ tons – that is to say, 2 followed by 27 zeros. This is equivalent to nearly 333,000 bodies the mass of the Earth, and a wastage of 4,000,000 tons per second is too slight to be noticed except over an immensely long period.

Eventually, of course, the supply of hydrogen 'fuel' will run low, and the Sun will change. It will rearrange itself, and new reactions will start. For a while the Sun will be much more luminous than it is today, and the Earth will certainly be destroyed, but there is no need to think about packing our bags and emigrating, because nothing dramatic will happen for at least 5,000 million years.

On the other hand, the Sun's output is not absolutely constant. In some respects it may even be regarded as a variable star, and there is a definite cycle of activity with maxima every 11 years or so. But some rather disquieting facts have come to light recently, and it seems that we know rather less about the Sun than we believed only a few years ago.

The solar surface is not always blank. There are darker patches to be seen, known as sunspots – rather misleadingly, because most of them appear to be hollows rather than elevations. Obviously they are not permanent, because permanent features cannot exist on a surface of seething gas, and no spot or spot-group lasts for more than a few months, while small ones often die out a few hours after they have been born. A typical spot-group consists of two members, a leader and a follower, together with smaller companions. With a large spot there is a central dark area or umbra, surrounded by a lighter penumbra; sometimes the shapes are regular, sometimes not. The umbra is at a temperature about 2000 °C. cooler than that of the surrounding bright surface or photosphere, which is why it appears black. If it could be seen shining on its own, its surface brilliance would be greater than that of an arc-lamp.

During a period of maximum sunspot activity there may be many spot-groups on view at the same time, while near minimum the disk may be clear for several consecutive weeks. The last maximum was that of 1980, so that the next may be expected about 1991, but we cannot be

Four photographs of the total solar eclipse of 31 July 1981. The maximum length of totality was 2 minutes 3 seconds. The track of totality crossed the North Pacific and also the Soviet Union; these four views were taken by Donald Trombino from the USSR. In order, the photographs show the beginning of the partial phase; the beginning of totality; mid-totality, with the corona well displayed; and the so-called 'diamond ring' effect as the first segment of the photosphere reappears from behind the Moon at the end of totality.

precise; the cycle is not completely regular, and the interval between successive maxima may be as much as 17 years or as little as 8 years.

Surrounding the photosphere is the chromosphere, a layer of much more tenuous gas, visible only when the Moon passes in front of the Sun and produces a total solar eclipse. Outside the chromosphere is the corona, or outer atmosphere of the Sun, which has no definite boundary, but simply thins out until the density is no greater than that of the interplanetary medium. During a total eclipse the corona is indeed magnificent. It is a pity that total eclipses as seen from one point on the Earth's surface are so uncommon; thus the last to be seen from England was in 1927, and the next will not be until 1999, when the track will pass over Cornwall. The Moon's disk is only just large enough to cover the Sun, so that the alignment has to be exact. Partial eclipses, when only a portion of the brilliant solar disk is hidden, are much less interesting, because the chromosphere and corona cannot be seen.

Before the Space Age, astronomers had to confine their studies of the outer corona to the fleeting moments of totality, and opportunities were limited, since the Sun can never be hidden for as much as eight minutes during any eclipse. One effort to overcome this difficulty was made at the eclipse of 1973, when totality lasted for over seven minutes. The view was superb; I saw it from a ship off the coast of North Africa and the corona was brilliant. The Moon's shadow sweeps across the Earth very rapidly, but a Concorde aircraft can fly fast enough to keep pace with it, and a Concorde was duly pressed into service; it took off from the Canary Islands at the appropriate moment, and kept beneath the shadow for 76 minutes, enabling the astronomers on board to carry out some very useful studies. Two holes were bored in the upper fuselage to accommodate the instruments. During the flight over Africa the Russians fired a high-altitude instrument-carrying rocket, and I was afraid that if it went off course there might be three holes in the Concorde, but fortunately all went well, though the record for totality has since been broken by the Skylab astronauts.

(There was one curious episode connected with the 1973 eclipse. One party of scientists went to Kenya, which was crossed by the track of totality, and was assured that the local inhabitants were friendly. As soon as the eclipse began, the nice, friendly natives surrounded the party and began to lob rocks, since they were under the impression that the white men had come to put out the Sun!).

Most of our knowledge of the Sun comes from instruments based on the principle of the spectroscope. When a beam of sunlight is passed through a glass prism or some equivalent device, the light is spread out into a coloured band. Light is a mixture of the colours of the rainbow, and the red part is bent or refracted less than the blue, so that we have a whole series of colours: red, orange, yellow, green, blue, and violet. This rainbow or 'continuous' spectrum is due to the Sun's bright surface. The chromosphere, however, does not yield a rainbow; instead we have disconnected lines, which would seem bright if they were shining on their own, but appear dark in the spectrum of the Sun, because they absorb certain wavelengths of the rainbow and are therefore called absorption lines. Each of them is due to a particular substance, and each

substance has its own particular trademark that cannot be duplicated. Thus two dark lines in the yellow part of the rainbow can be due only to the element sodium, and we can tell that there must be sodium in the Sun.

Spectroscopes, and other instruments based upon the same principle, allow us to study the chromosphere at any time, without waiting for an eclipse, and we can also see the prominences, glowing clouds of red hydrogen rising from the surface. The corona is more of a problem; fortunately we can now start to use space vehicles, from which the situation is much better. The first true American space station, Skylab, carried solar equipment, and the astronauts used it well.

Rather surprisingly, the corona has a temperature of well over

1,000,000 °C. Yet it is not 'hot'. Scientifically, temperature is defined by the rate at which the various atomic particles move around. In the corona these velocities are very high, and so is the temperature; but there are too few particles to give out much heat. The best comparison I can give is that of a firework sparkler and a red-hot poker. Each spark of the firework is at a high temperature, but it has so little mass that it is quite harmless – whereas I for one would not care to pick up the glowing end of the poker.

The Sun sends out streams of low-energy particles in all directions, making up what is called the solar wind; it was first studied in detail from space probes, notably the artificial satellite Explorer 10 in 1961, the early Sputniks, and Mariner 2 on its way to Venus. More recently there have been some elaborate spacecraft designed for solar research, the latest of which is the Solar Maximum Mission, launched in 1980. The solar wind rushes past the Earth at around 900,000 m.p.h., and is 'gusty'. At its strongest, particularly when coming out through calm areas of the corona known as coronal holes, it overloads the Van Allen belts, so that electrically charged particles cascade down into the upper atmosphere and produce brilliant displays of auroræ or polar lights. The solar wind also acts on the tails of comets, and pushes them outward, so that when a comet is receding from the Sun it actually travels tail-first.

Despite all our efforts, it cannot be said that we yet have a proper understanding of the cause of sunspots, but they are definitely the centres of strong magnetic fields, even though the overall field of the Sun is very weak. However, we have made marked progress during the last 20 years, and the basis of modern theory was worked out in 1961 by Horace Babcock at the Mount Wilson Observatory.

Launch of the Solar Maximum Mission probe from Cape Canaveral on 14 February 1980. The launch vehicle was a Delta rocket. The probe was designed specially to study solar flares; it carried seven scientific instruments, designed to study flares not only in ultra-violet but also in the X-ray and gamma-ray regions of the spectrum. It proved to be very successful, but serious faults developed during mid-1980; it began to 'wobble', tracing out a cone about its base as it moved along in orbit. The present plan is to send up an astronaut to repair it! In late 1983 it is hoped that a member of a Shuttle crew will go over to it, piloting a jet-powered backpack, and steady it; it will then be grappled by the Shuttle's remote manipulator arm, put into the cargo bay, repaired, and returned to orbit. If successful, this will be an historic 'first'!

Mount Wilson, overlooking the city of Los Angeles, was for many years the senior observatory of the world, because it had the largest telescope: the 100-inch Hooker reflector, completed in 1917. Not until the Palomar 200-inch began its career, more than 30 years later, was the Hooker reflector challenged; it was in a class of its own. But Mount Wilson Observatory was originally founded for studying the Sun, mainly through the efforts of George Ellery Hale, who had made fundamental advances in solar research when still a young man. Hale was persuasive as well as skilful, and he had the happy knack of persuading friendly millionaires to finance his schemes. No doubt this was much easier 80 years ago than it is now, but it was still a very considerable achievement, and Hale's efforts led to the setting-up of several major telescopes, including the Hooker reflector and, finally, the Palomar 200-inch.

The Mount Wilson story really began in 1904, with the completion of the Snow solar telescope, named in honour of the benefactor who provided the money for it. Old though it is, the telescope is still in excellent working order. The basic principle is the same as the one at Kitt Peak, but its focal length is much shorter (only 60 feet) and the tunnel is horizontal instead of being raised. Frankly it looks rather unimpressive, and, from the outside might easily be mistaken for a collection of garden sheds, but it has done good work over the years. When it was first installed, all the various parts had to be hauled up the mountain trail by mule-power, because in those days there was no road to the site.

The 60-inch reflector at Mount Wilson – the first of the Observatory's large telescopes, and still in full operation when I photographed it in December 1981.

The Snow telescope was so successful that two others were soon built, a 60-foot instrument in 1907 and a 150-foot in 1912. The only real modification was that the towers were vertical, so as to avoid disturbances due to the heated air at ground level. With the larger of the two, the tower is nearly 170 feet tall. At its top there are two mirrors, one of which is turned slowly by clockwork to follow the motion of the Sun. These mirrors reflect the light vertically downwards through a 12-inch diameter lens of 150 feet focal length, producing an image of the Sun 17 inches in diameter in the observation room. To reach the top of the tower involves using a somewhat precarious elevator cage, hydraulically powered. It has been known to stick, and on one occasion an astronomer was stranded half-way down, so that he had to climb back to terra firma by way of the girders – not something that I personally would like to attempt.

It was with the 150-foot tower telescope that Hale first discovered the magnetic fields of sunspots. The solar cycle was already well known, but could it be regarded as a permanent phenomenon? Or, alternatively, could the Sun be behaving abnormally at the present epoch? Even today we cannot be sure, and investigations into the solar cycle make up the main programme of research now being carried out at Mount Wilson by Dr Robert Howard and his colleagues.

According to Babcock's theory, the lines of magnetic force in the Sun lie below the bright surface. The Sun does not spin in the way that a solid body would do; the equator has a rotation period of 25 days, but in higher latitudes the period is several days longer, and therefore the magnetic field lines become twisted. Eventually they break through the bright surface, cooling it and producing sunspots. If there is a sort of loop, resulting in two spots, it is natural for one member of the pair to have 'north' magnetic polarity and the other 'south'; this is exactly what we find.

However, the sunspot cycle has not always been the same. A century ago the English astronomer Walter Maunder examined old records, and found that between 1650 and 1715 there were practically no sunspots at all, so that the cycle was in abeyance; this has become known as the Maunder Minimum (though to be fair, Maunder was not the first to draw attention to it; he had been anticipated by a German named Gustav Spörer). During this period, the weather over much of the world was exceptionally cold, and in England the Thames froze every winter. There is evidence of earlier spotless periods – for instance between 1400 and 1510, and between 640 and 710 – but the records are too incomplete to be trusted, whereas the Maunder Minimum is well authenticated. Evidently the Sun is not so steady and unchanging as might have been expected.

There are also some newly found short-term variations. In particular there are oscillations in the whole of the Sun, rather like the ringing of a bell. In 1975 Dr Henry Hill, in Arizona, set out to measure the amount by which the Sun is flattened at its poles, and found that the diameter showed a slight variation with a mean period of just over three-quarters of an hour. Subsequently, British and Russian observers independently reported a variation with a period of 2 hours 40 minutes, though the total change in the Sun's diameter is a mere 12 miles, not much when we remember that the full diameter is 865,000 miles.

There are also 'surface waves', much too slight to be seen visually, but detectable with the spectroscope. The waves may be only a few feet or a few yards high, but, unlike waves in our oceans, they extend deep into the solar globe. Vast areas of the Sun's surface rise and fall in periods of a few minutes. Theorists have constructed mathematical models to try to explain this behaviour, and have come to the conclusion that the Sun's interior is spinning round much more quickly than the outer layers, but no precise estimates can yet be made, because we do not know the size of the spinning core.

At Mount Wilson, Dr Howard and Dr Barry LaBonte have used the 150-foot tower telescope to study currents in the solar surface. They began their work in 1966, making accurate, regular measurements of east-west velocities at 24,000 points on the Sun's photosphere. By 1980 they were ready to announce their results. There are always two fast and two slower currents in each hemisphere, symmetrical to the equator, and over the course of a solar cycle the currents migrate towards it. The spots behave in the same way, as had been known for a long time. They are most numerous in the boundary zones between the fast and the slow currents, so that evidently there is a close link between the 11-year cycle and large-scale mass movements in the Sun's globe.

Longer-term changes cannot be studied so conveniently, because they take decades or even centuries to make themselves evident. It is generally believed that the Ice Ages on Earth, the last of which ended only 10,000 years ago, have been due to slight variations in solar output (though we cannot dismiss Sir Fred Hoyle's wandering asteroids), but we are not now confident that things are completely normal at the present moment. This brings me on to one of the most baffling of all the problems of modern solar research: the lack of neutrinos.

The production of energy in the Sun's core is by no means a straightforward process. Certainly it involves the transformation of hydrogen into helium, but in a rather roundabout way, involving several types of reactions. Particles known as neutrinos may be expected to be produced, and are absolutely fundamental to the whole process. Neutrinos are elusive things. They have no electrical charge, and no mass (or at least, very little; doubts about this have been expressed recently), so that they are hard to track down, particularly as they can pass unchecked through most materials. They can, for instance, go right through the Earth without being blocked or diverted.

However, there is one way to catch them. A neutrino may occasionally interact with the nucleus of an atom of chlorine (one of the two elements making up common salt), changing it into another kind of atom known as argon-37, which is radioactive, and therefore relatively easy to identify. This does not happen very often, but it gives investigators their best chance.

To detect neutrinos from the Sun, you need a great deal of chlorine. The most convenient way to provide this is to use a large tank filled with what is nothing more nor less than cleaning fluid, known technically as perchloroethylene. Its chemical formula is $C_2Cl_4$, which indicates that each molecule is made up of two carbon atoms together with four of chlorine. When an interaction occurs, argon-37 is produced. The numbers of argon-37 atoms formed can be measured, and this in turn gives the numbers of neutrinos that have scored direct hits.

Unfortunately, the experiment will not work at ground level, because cosmic-ray particles can produce exactly the same effects. The only solution is to 'go underground', and this brings me to the world's most unusual observatory: Homestake Mine, in the town of Lead in South Dakota. It is here, at a depth of 4850 feet, that Dr Raymond Davis and his colleagues from the Brookhaven National Laboratory have set up their experiment.

South Dakota has a romantic history. Lead (pronounced 'Leed', and nothing to do with the metal) lies a few miles from Deadwood, which is always associated not only with gold but also with the gunslingers of little over a century ago: Wild Bill Hickok, Calamity Jane, Dr John F. ('Doc') Holliday and the rest. Wild Bill was actually killed at a gunfight in a Deadwood saloon on 2 August 1876. Today the gunslingers have gone, but the gold is still there. Homestake was opened in 1877, and has remained in full production ever since. It is the largest mine in the United States, and I am told that it produces 0.7 of one per cent of all the gold in the world, but anyone going there in the hope of picking up a few stray nuggets will be disappointed, since it takes seven tons of rock to produce a single ounce of gold.

I first saw Homestake in January 1982, when I visited the underground observatory. The first stage is to dress up. You have to don a hard helmet, together with safety spectacles, a miner's lamp, and – of special importance – a device known as a self-rescuer. Should there be a sudden accumulation of the deadly, odourless gas carbon monoxide, you can unfasten the flask-sized rescuer, put the nozzle in your mouth, clip your

The 'control room' of the underground solar observatory at Homestake Mine, with Dr Raymond Davis, who designed the experiment. The gas from the tank comes into the control room via the large cylinder to the left, while the analytical equipment is to the right. A photograph which I took in January 1982.

nose, and breathe; provided that you fix the rescuer properly in position, it can give you an hour's grace.

I went down in the miner's cage (a distinctly rough ride, I may add!) and then made my way along the tunnel leading to the observatory, making sure to keep clear of the overhead cable, which carries enough electricity to fry you like an egg should you happen to touch it. After a few minutes' walk I came to the observatory, which was pleasantly warm (around 80 °F.) even though the temperature above was some 20 degrees below freezing-point, and the Black Hills were covered with deep snow, as indeed they are for several months in each year. There are two main sections: the control room, containing the analytical equipment, and the detector room, containing a huge tank that holds no less than 100,000 gallons of cleaning fluid – and, hence, a great many atoms of chlorine. This tank is the observatory's 'telescope'. Only two or three cosmic-ray particles per square metre per day can pass through the mile of rock overhead, whereas the massless, chargeless neutrinos have no trouble at all.

Dr Davis and his colleague, Dr Keith Rowley, showed me round. The tank itself is surrounded by water, and at present it is covered up by detectors making up a different experiment, designed to trap the few cosmic-ray particles that do manage to get through, but it is impressive by any standards.

The tank is then left for a matter of eight weeks or so, to see how many atoms of radioactive argon-37 have been produced. Over 70 'runs' have been completed since the experiment began in 1965, and the results have been pleasingly consistent.

Obviously the experiment is very delicate and difficult. Over an eight-week period, perhaps ten atoms of argon-37 are produced, whereas the whole tank contains about a thousand million million million atoms of various types, so that the greatest care must be taken. The fluid in the tank is mixed by pumping helium into it, and the gases are then fed through into the control room, where they are caught in a vessel that contains charcoal. The argon-37 sticks to the charcoal, and when the charcoal is subsequently heated the argon is given off. It passes through what is called a titanium oven, designed to remove the last traces of unwanted gases, then into a second charcoal trap, and finally into what Dr Rowley described as 'the eyepiece of the telescope'. This is taken out, and put into a large lead box thick enough to shield it from all other radiations, so that the counts of the radioactive argon-37 atoms can come through on a ticker-tape.

The results have been startling. Theory and observation do not agree; the Sun is sending out only about a quarter as many neutrinos as it had been expected to do. Dr Davis had anticipated registering at least one per day, but the actual flux is very much less than this.

Naturally enough, I asked whether there could be any fault in the experiment. After all, tracking down ten special atoms in so many thousands of millions is no easy matter. But exhaustive checks have been made; for instance, the fluid in the tank has been deliberately seeded with a few argon-37 atoms and the usual tests have been made, with the advantage of knowing what the answer should be. It always works. There

My picture of the tank at the neutrino observatory – now covered over by the 'boxes' concerned with the separate cosmic-ray experiment.

seems no reasonable doubt that the equipment really is giving a faithful count of the numbers of solar neutrinos.

If so, then there is something wrong with our theories about the Sun itself. One possibility is that the temperature at the solar core may be less than is generally believed. Neutrino production is very sensitive to temperature, and if we reduce the value at the Sun's centre to 14,500,000 °C. instead of the accepted 15,500,000 °C. we can explain the paucity of neutrinos, but this will cause all sorts of theoretical headaches in other directions. There may also be errors in our estimates of the extent to which material in the Sun's globe is being mixed. Yet another possibility is that neutrinos may break up during their long journey from the Sun to the Earth, but this also raises problems, and physicists tend to be sceptical.

Finally, it is just conceivable that the Sun may be behaving in an abnormal way at the present time, and we know that it is certainly variable to some extent, though the neutrino results are not yet sufficiently clear-cut to tell whether or not there is any connection with the eleven-year cycle. Similar experiments are being undertaken, in the Soviet Union particularly, and even more sensitive tests are being worked out, but in any case the Homestake results have provided abundant food for thought.

The real importance of the work is that it gives us our only hope of studying particles which have come more or less directly from the Sun's core. In learning more about the Sun, we are also learning more about

In the control room at Homestake Mine; Dr Raymond Davis and Dr Keith Rowley with the container in which the sample from the main tank is placed after collection.

conditions on other stars, and about particle physics in general.

Before we can be confident that we really know how the Sun behaves, the neutrino problem must be solved; but at present the mystery remains, and Homestake Mine provides us with our best chance of tackling it.

The idea of an inconstant, shaking, wobbling Sun is still novel, but you will see now why I commented that our knowledge is much less than we fondly imagined even a decade ago. We must also consider the Sun's other invisible radiations, from radio waves at the long-wave end of the spectrum through to the ultra-violet, X-rays and gamma-rays at the short-wave end. Another important method of attack is to study other solar-type stars, and see whether they show cycles of activity similar to the Sun's. Here the pioneer work has been carried out at Mount Wilson by Dr Olin Wilson and his colleagues.

The initial difficulty is that a star looks like nothing more than a speck of light, so that the only way to approach matters is to use the spectroscope. For this kind of work the tremendous light-grasp of the 100-inch Hooker reflector is not essential, and the second of the Mount Wilson telescopes, the 60-inch, has been pressed into service. At the moment, this programme monopolizes all its time.

Solar-type stars, of which there are plenty in the Galaxy, show spectra which are very similar to that of the Sun, with the usual rainbow background crossed by dark absorption lines. Two prominent lines, lettered H and K, are due to the element calcium. During periods of great surface activity, the H and K lines become bright instead of dark, because of the rising, intensely hot gases. Therefore, if the spectrum of a solar-type star shows bright H and K lines, we know that tremendous activity is taking place. The Sun has strong convection currents in its outer portions – that is to say, heat is being carried upwards by the actual movement of material – and it is reasonable to assume that the same is true for other stars.

At Mount Wilson, A. H. Vaughan built special equipment to use on the 60-inch reflector. The results were encouraging. Definite variations in the spectra were found, and by the time that over 90 stars had been examined, it was possible to put forward some suggestions which were backed up by an impressive amount of observational evidence. The work was so important that during 1980 the observatory director, Dr G. W. Preston, contacted other observatories with a request to take part in a co-operative programme. All large telescopes are over-subscribed – that is to say, there are not enough of them to go round, so that observers anxious to undertake specialized researches have to take their places in the queue. In this case the observers pooled their time for 100 nights, and the 60-inch was used for this particular project alone.

Some stars were found to show cyclic behaviour. Moreover the H and K lines showed regular, short-term changes, and it seemed that activity on their surfaces was limited to one hemisphere only. When the active or 'spotted' side was facing us, the H and K lines were bright; when the active side was turned away from Earth, the lines became dark, and therefore it was possible to measure the rotation periods of the stars concerned. With other stars there seemed to be continuous chromospheric activity, sometimes thirty times as energetic as with the Sun.

Some stellar rotation periods had been measured much earlier, in a different way. If a light-source is approaching us, the wavelength is apparently shortened, and the light is slightly bluer than it would otherwise be; if the source is receding, the wavelength is lengthened, and thc light is reddened. (This is the celebrated Doppler Effect. It is also noticeable with sound-waves, as you will find if you listen to a police car or an ambulance passing by with its siren blaring; the note of the siren drops when the vehicle starts to move away.) With the spectrum of a star, or any other light-source, the dark lines will be accordingly shifted either to the red or the blue, which is how we measure towards-or-away velocities. As a star rotates, one edge of it will be approaching the Earth and the other receding, so that the dark lines are broadened, and the rate of spin can be found. This effect is naturally superimposed on the overall movement of the star, but in many cases it is detectable. However, it is of use only with the quick-spinning stars, and stars like the Sun rotate much more slowly.

By studying other stars we can learn more about the Sun itself, and the converse is also true; by now it has become impossible to separate astronomy out into distinct, well-marked departments. Mount Wilson deals with all aspects. It is one of the oldest of the great observatories, but it has shown itself able to move with the times, and even though it no longer owns the world's largest telescope it is still very much in the forefront of modern research.

# 11 New Techniques

Twenty-five years ago there were two existing telescopes of outstanding size: the 100-inch Hooker reflector at Mount Wilson and the 200-inch Hale reflector at Palomar. They were 'out on their own' and had no close rivals; moreover each had been responsible for major advances in astronomical science. For example, the 100-inch had been used to show that our star system or Galaxy is only one of many, while the 200-inch had demonstrated that the universe is twice as large as had previously been thought. At this time, too, virtually all observation was carried out by means of photography. Radio astronomy was in its infancy, and infra-red astronomy even more so, while short-wave astronomy had not even begun.

The situation today is completely different. Great new telescopes have been built, and the older ones have been given new leases of life by the introduction of new recording equipment; photography is fast being superseded by electronic devices. We are able to study a wide range of the total electromagnetic spectrum, and observations are being made from above the top of the Earth's atmosphere, both by automatic satellites and probes, and by manned spacecraft. I suppose that more progress in astronomy has been made since 1957 than at any comparable period in history – and I am not thinking only of the Apollo flights to the Moon.

One difficulty about writing a book of the present kind is that all the branches of astronomy are intertwined. Therefore I propose to write a somewhat rambling chapter, and say a little about some of the new observatories and equipment.

I must, I suppose, take the Palomar reflector as my starting-point, because it has been so drastically modified since it was completed in 1948. As I have said, the observer no longer sits in his cage within the telescope, but controls everything from a comfortable distance and watches the results displayed on his television screen. Moreover, the electronic equipment now in use is far more sensitive than any photographic plate, so that the range of the 200-inch is extended. It may be a quarter of a century old, but it is still larger than any other telescope in the Western world, and for sheer size it is inferior only to the newer telescope which the Soviet authorities have set up in the Caucasus Mountains.

The Palomar 200-inch reflector, as I photographed it in 1979.

It was natural enough for the Russians to consider building a really giant telescope. By 1961 they had completed a 102-inch reflector, and set it up at the Crimean Astrophysical Observatory; it is an excellent instrument, though seeing conditions in the Crimea are not so good as in, say, California or Chile. Soon afterwards the Soviet Academy of Sciences announced the intention of building a 236-inch reflector on Mount Semirodriki, not far from the Zelenchukskaya station in the northern Caucasus. It is reasonably high (well over 6000 feet) and the seeing is probably as good as any in the less inaccessible parts of the U.S.S.R., though it must be added that anyone going to Semirodriki except near midsummer must be prepared for long delays.

The dome itself is conventional enough; it weighs 1000 tons, and was made in Leningrad so that it had to be taken to Semirodriki in pieces. The mirror, made in Moscow, weighs 42 tons, and this presents problems, because any piece of glass as heavy as this tends to distort under its own weight when it is moved around. The Russians, therefore decided on a daring plan. Instead of the conventional type of mounting, they would use something that seemed at first to be going backwards in time. They would abandon the equatorial kind of mounting, and use the old-fashioned altazimuth system.

Because of the Earth's rotation, any celestial body will move across the sky in an east-west direction, and its altitude above the horizon will also change constantly. If you turn a telescope towards a star and leave it motionless, the star will race out of the field of view. To compensate for this, all large telescopes are mechanically driven, and they are mounted upon an axis that is parallel to the Earth's axis of rotation, so that the up-or-down motion looks after itself, and only one driving motor is

Dome of the Russian 236-inch reflector at Zelenchukskaya, Caucasus.

needed (generally electrical, though the Mount Wilson 100-inch is still driven by falling weights; the results are so good that there has never been any need to alter the system). On the other hand a small portable telescope can move freely in any direction, either in *alt*itude or *azimuth* – which is why such a mounting is termed an altazimuth.

Before computers came upon the scene, there was no choice but to use the equatorial type of mounting for a large powerful telescope, but by the 1960s it had become possible to drive a telescope in both senses simultaneously – up or down, and east to west. This would mean that the mirror would not have to be moved to so great an extent and would be less liable to distortion.

The Russian plan worked well. The 236-inch was ready for use in 1976, and the mounting at least was a success. The optics have been more of a problem, and it cannot honestly be said that the results coming from the observatory have yet been as good as expected, but there is plenty of time, and a giant instrument such as this is bound to have prolonged teething troubles.

Whether any telescope larger than the 236-inch will ever be set up on the Earth's surface is by no means certain, because each increase in aperture adds to the difficulties caused by the dirty, unsteady atmosphere and on the whole it is probably more valuable scientifically to concentrate on building more telescopes in the 100- to 200-inch range. Also, making a huge mirror is not easy, and maintaining it in first-class condition is a tremendous task. However, suppose that we could use several mirrors of more modest size, and combine them to produce a single image?

This is just what has been done with the MMT or Multiple-Mirror Telescope, high in the Santa Rita Mountains of Arizona. The building perches on the very top of Mount Hopkins, between 8000 and 9000 feet above sea-level, and when seen from below it looks almost like a child's toy; the flat area on the mountain-top is only just big enough to accommodate the building and essential parking spaces. The drive up the mountain is distinctly hazardous. The road is rough – little more than a track – and is narrow and winding; to one side is a rock wall, and to the other a sheer drop of hundreds of feet. Neither is there a safety-rail. I have negotiated it only in good weather, but I am told that during the season of ice and snow the road becomes very interesting indeed.

The housing of the MMT is absolutely unlike the graceful domed structures of Mount Wilson or Palomar. It is square, and the whole building moves round with the telescope. The rotation is extremely smooth, even though the building weighs 500 tons, but there are characteristic problems; once, for instance, the rails were choked with swarms of ladybirds, which had to be painstakingly scooped out. (Subsequently special sweepers were introduced, able to cope not only with ladybirds but also with other hazards such as dropped sandwiches.)

Instead of one large mirror, the MMT uses six. Each is 72 inches in diameter, and of 'eggcrate' construction, weighing only about one-third as much as a conventional mirror of the same size. They work together; the light they collect is brought to the same focus, and is combined. The overall result is equal to a reflector with a single 176-inch mirror,

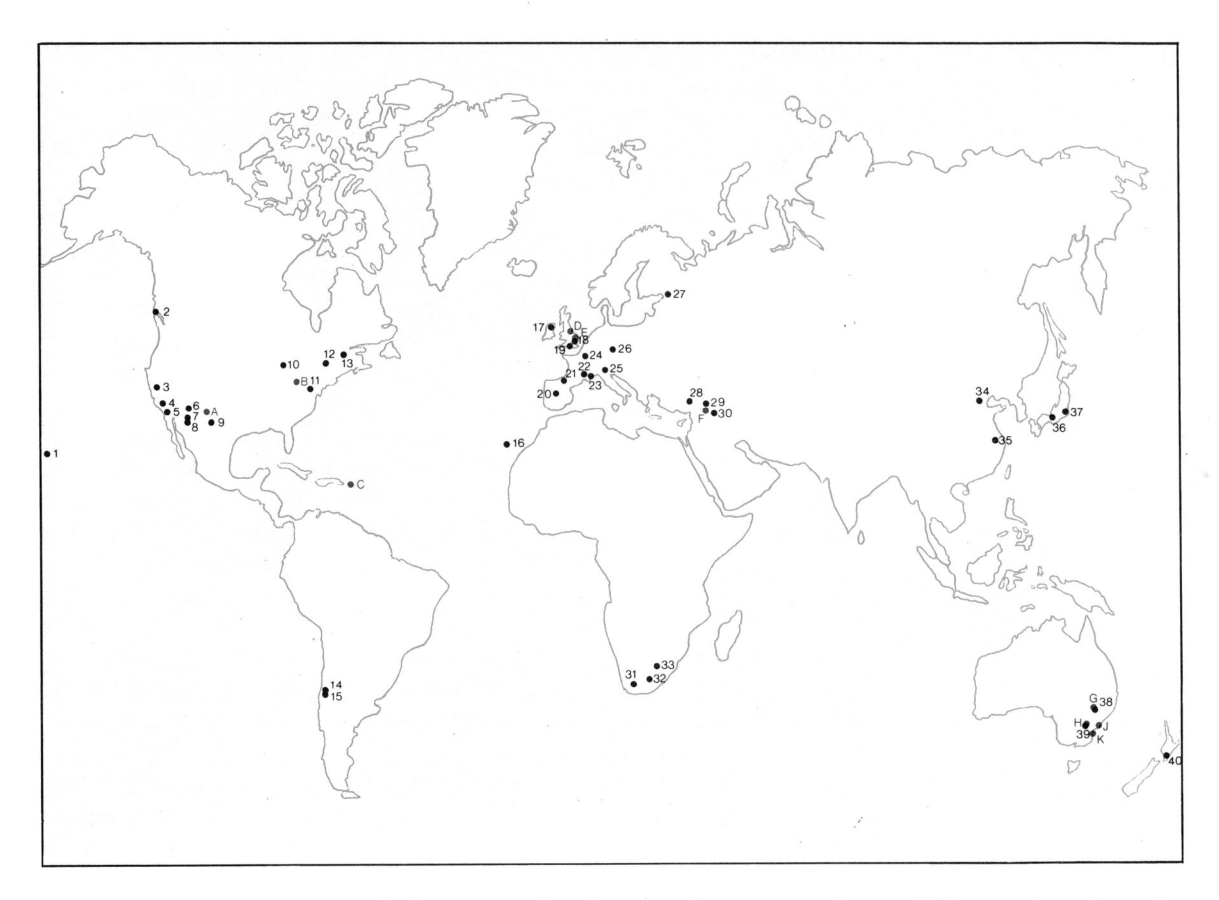


**Observatories throughout the world.** Observatories that are mainly optical are shown in black. Radio observatories are shown in red.

OPTICAL

1 Mauna Kea, Hawaii
2 Dominion Astrophysical, Victoria, Canada
3 Lick, Mt Hamilton, California
4 Mt Wilson, California
5 Mt Palomar, California
6 Flagstaff, Lowell Observatory, Arizona
7 Kitt Peak, Tucson, Arizona
8 MMT, Mt Hopkins, Arizona
9 McDonald, Texas
10 Yerkes, Williams Bay, Wisconsin
11 Royal Naval, Washington DC

and there are still only two telescopes in the world larger than that: the Russian 236-inch and the Hale reflector at Palomar.

The MMT does not even look like a telescope. It is short and squat, with a maze of girders and wires, in the midst of which are the six 72-inch primaries. There is also a separate 30-inch mirror which is not connected to the main system, but which acts as a guide. All the mirrors are silica-coated, and are just as accurate as those of any other telescope.

Keeping even a few mirrors in proper alignment is tricky enough, and the problems are enormously increased with the MMT, since if any one of the six primaries is even slightly out of position the telescope will not perform properly. The method adopted is to use a laser. A laser beam may be described as a pencil-thin shaft of very intense light – known technically as coherent light – which does not spread out, and can therefore be very accurately positioned. In the MMT, a transmitter sends a laser beam through the optical system, producing what is to all intents and purposes an artificial star. The required position of this artificial star is known, and if the telescope is producing an image exactly where it ought to be, then everything is properly aligned; if not, an automatic adjustment is made. The laser control has to be kept in operation all the time that observations are being made, because conditions alter whenever the telescope is shifted; gravitational strains, for example, are quite enough to produce unacceptable distortions of the image unless they are corrected.

12 David Dunlap, Toronto, Canada
13 Mt Megantic, Quebec, Canada
14 Cerro Tololo, Chile
15 European Southern, La Silla, Chile
16 Tenerife, Canary Islands
17 Armagh, N. Ireland
18 Old Greenwich, England
19 Herstmonceux Castle, Hailsham, England
20 Madrid, Spain
21 Pic du Midi, Pyrénées, France
22 Haute Provence, France
23 Nice, France
24 Meudon, France
25 Milan and Turin, Italy
26 Ondrejev, Czechoslovakia
27 Pulkova, Leningrad, USSR
28 Crimean Astrophysical, USSR
29 Zelenchukskaya, USSR (236 in)
30 Byurakan, USSR
31 Sutherland, S. Africa
32 Boyden, S. Africa
33 Johannesburg, S. Africa
34 Peking, China
35 Purple Mountain, China
36 Kyoto, Japan
37 Tokyo, Japan
38 Siding Spring, New South Wales, Australia
39 Mt Stromlo, ACT, Australia
40 Carter, Wellington, New Zealand

RADIO

A VLA, Socorro, New Mexico
B Greenbank, W. Virginia
C Arecibo, Puerto Rico
D Jodrell Bank, England
E Cambridge, England
F RATAN 600, Zelenchukskaya, USSR
G Culqoora and Narrabri, New South Wales, Australia
H Parkes, New South Wales, Australia
J Mills Cross, Fleurs, New South Wales, Australia
K Molonglo, Australia

Diagram by Paul Doherty

In theory this method is excellent. Unfortunately it has given some trouble, because the telescope is open to the air as soon as the slit in the 'dome' has been drawn back, and insects flying about in the laser beam have proved to be a major hazard. Experiments are still going on, and no doubt a solution will be found before long, but even without the laser control, the MMT works excellently, and astronomers who have made observations with it are unanimous in saying that it is just as convenient and just as powerful as a telescope built in a more orthodox pattern. It cost much less than a single 176-inch mirror would have done, and it may be the first of a whole new generation of multiple-mirror telescopes. The mounting, of course, is altazimuth, and computers control the driving mechanisms without any trouble at all.

Quite apart from the technical problems of telescopes themselves, there is the need to choose good sites for them, and in earlier times the importance of this was not fully appreciated. England provides the classic example. The original Greenwich Observatory was built in the Royal Park, and when this became hopeless as an astronomical site the main instruments were shifted to Herstmonceux in Sussex; the old castle there was taken over as the main headquarters, and in the grounds various domes sprouted like mushrooms. In 1967 came the Isaac Newton Telescope or INT, the largest ever made in the British Isles. It was a reflector with a 98-inch mirror, conventionally mounted, and used in conjunction with all the new electronic aids.

Unfortunately, one has to admit that the British climate is not at all to the astronomer's satisfaction. There is far too much cloud, and at Herstmonceux there was the added complication that the lights of the nearby town of Eastbourne soon started to become obtrusive. After much deliberation it was decided to shift the INT to a better home, and the choice fell upon La Palma in the Canary Isles. A new observatory was planned. It was originally called the Northern Hemisphere Observatory, but is now known by the more dignified title of El Observatorio del Roque de los Muchachos. Los Muchachos, or 'Boys', are twin rocks on the highest point on the island. I am sure that there must be a local legend about them, but nobody at La Palma has ever been able to tell me what it is!

'The Boys' lie at an altitude of 7400 feet, and the Observatory is nearly as high. The climate of the Canaries is good, and there was already an observatory on the neighbouring island of Tenerife, so that exhaustive tests had already been made. The islands are Spanish, and the Observatory is officially Spanish, but it has been organized on a fully international basis, with Holland, Sweden, Denmark and other countries involved as well as Britain. There are the closest possible links with the Royal Greenwich Observatory at Herstmonceux, and the first Project Scientist at La Palma, Dr Paul Murdin, comes from Greenwich, while the present Director of the Royal Greenwich Observatory, Professor Alec Boksenberg, is in overall charge.

The Observatory was formally opened in May 1979, by which time building operations were well under way. There will be three main telescopes. The largest will be a new 180-inch reflector, the William Herschel Telescope, with an altazimuth mounting much on the lines of

The MMT or Multiple-Mirror Telescope at Mount Hopkins, as I photographed it in 1981. The MMT itself is visible inside the observatory; the whole observatory turns with the telescope, so that the opening is always in the correct position. On the flat top of the peak there is just about enough room for the observatory and a small car park!

the 236-inch at Semirodriki; the INT, equipped with a new mirror 102 inches in diameter, will be the second telescope, and there will also be a 40-inch reflector, while the Swedes already have a solar observatory in working order.

In the summer of 1981, when I went to La Palma, the scene was one of tremendous activity, and the INT was already in place inside its dome, though not yet operational. The site is impressive, even though the road up the mountain is very rough; twice our Land Rover was stuck in the mud. Great things may be expected from La Palma. In my view, one of the most important factors is the international one; astronomy knows no boundaries, and there are no Iron Curtains among the stars.

The only real disadvantage of La Palma is that since it lies well to the north of the equator, the telescopes cannot see the skies of the far south, where we find some of the most interesting objects in the sky – notably the two Clouds of Magellan, the nearest of the really large galaxies beyond the Milky Way. We need telescopes in the southern hemisphere as well, and today the requirements are being met. I will have more to say later about Siding Spring Observatory in Australia, and I will pass over the three giant telescopes in Chile – La Silla, Cerro Tololo and Las Campanas – because, to my regret, I have never been there. The first two observatories have reflectors of over 150 inches aperture, and Las Campanas, officially allied with Mount Wilson in California, has the Irénée du Pont 100-inch.

South Africa, too, is well in the picture. Most of its main telescopes, including a 74-inch reflector, have been assembled at Sutherland in Cape Province, where conditions are particularly good; there is also a 60-inch at the Boyden Observatory in Bloemfontein. Another of the Boyden domes, housing a 13-inch refractor, has an unwelcome occupant in the form of a cobra, which lives below the floorboards. It has never appeared during my spells there, for which I am duly grateful, but it is on record that a visiting astronomer entering the dome in the late evening was greeted by a ten-foot snake, after which he rather naturally decided to abandon observing for the night.

Finally, in this admittedly rambling survey, I must say something about another Boyden telescope, because even though it is small and old it has played a major role in astronomical history. It is a refractor with a modest 10-inch object-glass, so that its light-grasp is limited, but it is decidedly useful, and it was originally set up at Arequipa in Peru, outstation of the Harvard College Observatory. Photographs of the southern skies were taken with it, and some of these covered the Clouds of Magellan, which look almost like broken-off parts of the Milky Way; European astronomers never cease to bemoan the fact that they lie so close to the south celestial pole.

At that time – the early 1900s – the nature of the Clouds was not known, and neither was it realized that some of the nebulæ – misty-looking patches in the night sky – are independent galaxies. Most astronomers tacitly assumed that our Milky Way system was the only one, containing all the stars in the observable universe.

The 10-inch Metcalf telescope at Boyden Observatory, South Africa, formerly at Arequipa in Peru.
The observatory is of the run-off roof variety, and the mounting is of the German type, using a counterweight to balance the weight of the telescope – a favourite pattern with both refractors and reflectors.

The INT or Isaac Newton Telescope. When I took this photograph, in October 1981, the telescope had been moved to its new home at La Palma, but was not yet operational.

When the Arequipa plates were sent back to Harvard, they were closely examined by Henrietta Leavitt, one of America's most celebrated woman astronomers. Many stars were shown in the Clouds, and some of them were variable in light. A variable star is just what its name implies; instead of shining steadily it brightens and fades over a period which may be a few hours, a few days, or even weeks or years. Among these variables are the Cepheids, named after the best-known member of the class, Delta Cephei in the northern sky. Cepheids pulsate, and they are as regular as clockwork, so that one always knows what they are going to do next. Also they are very luminous, so that they can be seen over great distances.

Before long Miss Leavitt noticed something very curious. There were plenty of Cepheids in the Small Cloud, and it seemed that those of longer period were brighter than those of shorter period. There was a definite law: the longer the period, the brighter the star. To all intents and purposes the Cepheids in the Small Cloud could be regarded as lying at the same distance from us, just as it is good enough to say, for example, that Westminster Bridge and Trafalgar Square are at the same distance from New York. It followed, therefore, that the longer-period Cepheids really were the more powerful, and the same would presumably hold good for the non-Cloud Cepheids closer to us.

Most important of all, this gave a clue to the distances of the stars. When it was known how luminous they really were, the distances could be worked out – allowing for complications such as the absorption of light in space. The Cepheids acted as 'standard candles', and their distances could be found merely by following their brightness variations over a few nights.

It would be hard to over-estimate the importance of this Cepheid Period-Luminosity Law. Within a few years Harlow Shapley, at Harvard, had identified Cepheids in globular clusters – regular systems of stars lying out towards the edge of the Galaxy – and had been able to measure the size of the Galaxy itself, now known to be about 100,000 light-years in diameter.

By 1917 the Mount Wilson 100-inch reflector was ready for use. Six years later Edwin Hubble was able to detect Cepheids in the 'starry nebulæ', and to show that they were much too remote to be included in the Milky Way. Finally, in 1952, Walter Baade used the Palomar 200-inch telescope to prove that because of an error in the Cepheid scale, the

The Large Magellanic Cloud, photographed with the UK Schmidt telescope at Siding Spring. This colour picture enables one to distinguish at a glance the older, yellowish stars, the young, hot blue stars and clusters, and the reddish nebulæ of glowing interstellar gas excited by the hottest and most luminous stars of all. The most extensive of these nebulæ, the mass of loops and filaments, is known as 30 Doradûs, also known as the Tarantula Nebula; it is seen to the right of centre, and is one of the largest of its kind known anywhere in the universe.

universe must be even larger than had been thought. The Clouds of Magellan are over 150,000 light-years away, and most of the other galaxies lie at distances of many millions of light-years.

All this is ancient history by now, but it does show the importance of studying objects in all parts of the sky. No doubt this discovery would have been made sooner or later, but it was the work carried out with the 10-inch refractor at Arequipa that gave the key. I have the greatest respect for this little telescope, now installed in its latest home in South Africa and still as good as it used to be in its Peruvian days over half a century ago.

The Small Magellanic Cloud, also photographed with the UK Schmidt telescope at Siding Spring. At a distance of about 200,000 light-years it is slightly further away than the Large Cloud; it too contains young and old stars, dust clouds, star clusters and glowing nebulæ. The two Clouds are actually joined by an invisible cloud of hydrogen gas. They may be regarded as satellite systems of our own Galaxy.

# 12 Stellar Births and Stellar Deaths

Have you ever been to Hawaii? If not, you probably picture it to be a sort of earthly paradise, with blue lagoons, glorious beaches and perpetual sunshine. For some of the islands this is not so very wide of the mark, but there is another side to Hawaii too, and some of the scenery is as grim and forbidding as anything to be found in the world.

Big Island, the largest of the group, has one fair-sized town, Hilo, and several smaller ones, but it is dominated by the lofty peaks of Mauna Loa (Long Mountain) and Mauna Kea (White Mountain). Mauna Loa is an active volcano, and periodically makes its presence felt in no uncertain manner; once, in 1881, lava from it came within striking distance of Hilo, and according to legend was halted only by the intervention of a local witch-doctor. Flanking it is Kilauea, one of the most constantly active volcanoes known. It has a gigantic caldera, Halemaumau, which erupts frequently, and periodically pours lava over one of Big Island's thoroughfares, the Chain of Craters Road. Even when quiescent (as it was when I went there in December 1981), the smell of sulphur and the jets of steam issuing from the ground give it a sinister aspect.

Mauna Kea, slightly higher and also a shield volcano, is extinct. At least, I hope so, because one of the world's major observatories has been built on top of it. There have been no eruptions for at least 3000 years, and Mauna Kea has probably passed the end of its active career, though periodical ground tremors do occur. The summit is covered with snow for much of the year, which has led to its name of White Mountain.

The observatory is 13,780 feet above sea-level. At such altitudes one is above 40 per cent of the atmosphere, and the air is extremely thin by normal standards, so that it can prove dangerous. Anyone going to the summit is advised to have a medical check up first, while to go straight up and then run up or down a flight of steps is to invite trouble. Different people are variously affected; despite my age (58 at the time) and my heavy build I had no trouble at all, but others did, and it is more or less essential to pause at the 'half-way house', Hale Pohaku, for at least a few hours before driving up the remaining 4000 feet.

The idea of setting up an observatory atop Mauna Kea originated with Gerard Kuiper, who reasoned – quite correctly – that it would be advantageous to go as high as possible in order to be free of the troublesome, turbulent lower atmosphere. He climbed the volcano, and made some tests, after which he managed to interest the University of Hawaii and other institutions. There was a great deal of scepticism; the crest of

The dome of the UKIRT on top of Mauna Kea; I photographed it in December 1981.

Mauna Kea was inaccessible because there were no roads, and working at such an altitude was a real problem both mentally and physically, because one's intake of oxygen is so much below normal.

Nevertheless, the idea was taken up. NASA became interested; so did the Royal Observatory at Edinburgh in Scotland. The final result (unhappily, postponed until after Kuiper's death) was the present-day Mauna Kea Observatory.

At the moment there are four main telescopes, and others are being planned. One of them, a 144-inch reflector, is operated jointly by the Canadians, the French and the University of Hawaii; it is a conventional telescope, and one of the world's best. The University of Hawaii has an 88-inch reflector. The two remaining large telescopes are designed mainly for use at infra-red wavelengths. One, operated by NASA, has a 120-inch mirror; the other is UKIRT (United Kingdom Infra-Red Telescope), which has an aperture of 150 inches, and is the largest of its kind ever built.

The design was completed in 1975, and the optics were made at Newcastle in England, while most of the rest of the telescope came from the firm of Hadfield in Sheffield. By October 1979 everything was complete and installed on Mauna Kea, so that the telescope was officially commissioned. It has a relatively small dome, and the UKIRT itself is unconventional inasmuch as it is comparatively lightweight in construction, with a mirror (made of low-expansion ceramic material called Cervit) much thinner than normal, weighing only $6\frac{1}{2}$ tons instead of the usual 16 or 17 tons. The UKIRT costs much less than an ordinary telescope, and it was, frankly, something of a gamble. Until it had been tested fully, nobody was quite sure how it would behave.

The reason for this apparently fragile design was that UKIRT was designed for use in the infra-red part of the spectrum, and these waves are longer than waves of visible light, so that there is no need to have so accurate a mirror. In the design of the UKIRT, it was reasoned that something less than perfect would serve, and the reduction in cost and ease of construction were important factors. In a way, the results were much better than had been expected. The UKIRT is so good that it can be used for ordinary observation as well as infra-red, which is a sheer bonus.

In appearance there is nothing to single out the UKIRT from a conventional reflector. There is the usual skeleton tube with equatorial mounting; the optical system is the Cassegrain, in which the incoming rays are reflected back up the tube and then sent to a focus by way of a hole in the main mirror. Yet there are some unusual features. Because of the frequent ground tremors, big enough to be classed as earthquakes, the design incorporates brass pins which act as mechanical 'fuses'. The effectiveness of this method was demonstrated in March 1979, when an earthquake affected the whole of Big Island, including Mauna Kea; the pins sheared, allowing the telescope to ride out the disturbance unharmed. A second, more violent earthquake six months later was also safely negotiated.

Observing from a height of nearly 14,000 feet has marked advantages. There is practically no scattered light; the air is steady, and, even more importantly, very dry, since 90 per cent of the atmospheric water vapour lies below. It is water vapour which is the infra-red astronomer's worst

The UKIRT itself; my photograph, December 1981. Seen here it looks just like a conventional reflector on a conventional mounting; the lightweight construction is not apparent.

enemy, because it absorbs the long-wave radiations coming from space.

The main problems of Mauna Kea are purely human. Nobody sleeps at the Observatory, and after a night or day spent at the summit there is a prompt and enforced return to the 'half-way house' of Hale Pohaku, below the critical level of 10,000 feet. There are also occasional storms, which can blow up in a matter of a few minutes and can make life very uncomfortable. Once, in January 1980, three observers were caught unawares, and when they tried to drive back to Hale Pohaku their Land Rover broke down, so that they had to return to the UKIRT dome for shelter. By then an 80 m.p.h. blizzard was blowing, and the road became impassable; to make matters worse the electric power failed, leaving the marooned trio without light, heat or means of communication. It was not until the fourth day that a rescue party managed to get through, by which time the astronomers had had to cut up any available wood to make a fire. Their diet had consisted exclusively of soup and tinned beans.

When I arrived at Mauna Kea the weather was very different; the sun shone brilliantly, and the temperature was pleasantly warm. The drive from Hilo to Hale Pohaku is easy, but the final ascent to the domes is more difficult, and only four-wheel-drive vehicles can tackle it with any confidence. Actually, the time taken to reach the summit from Hale Pohaku is no more than half an hour, but one becomes very conscious of the thinning air, and any physical exertion leads to breathlessness and a thumping heart.

As you round the last bend in the dirt road, the domes come into view very suddenly. First there is the UKIRT; then the 88-inch and the Canada–France–Hawaii 144-inch, with the NASA dome further away and slightly lower down. To the right-hand side of the road lies the actual summit of Mauna Kea, while to the left there is a pair of peaks separated by a valley. This is generally known as Sub-Millimetre Valley, because it is hoped to set up new telescopes there operating in the sub-millimetre range of the spectrum – that is to say, radiations whose wavelengths are less than one millimetre but still in the infra-red range, too long to affect one's eyes.

Since infra-red light is invisible, it has to be studied by means of special equipment. I will not attempt to go into detail, because that would taken many pages; I will only say that detectors have to be kept extremely cold, which is done by surrounding them with liquid nitrogen or even liquid helium.

Most of the infra-red sources in the sky are weak, and care must be taken to separate them from the natural 'infra-red brightness' of the sky. This is done by a process known as chopping. The secondary mirror of the telescope can be rocked to and fro, so that the radiations come first from the empty sky and then from the region containing the source. The difference in signal strength between the two then gives the real strength of the source. With the UKIRT, the secondary mirror is rocked ten times every second.

Work has been continuous ever since the UKIRT came into operation. For the moment let me deal exclusively with infra-red studies, because the others are of the conventional sort. First, then, what can this invisible 'light' tell us? To answer this, it will help to consider the Orion Nebula.

Nebulosity in Orion, including the characteristic dark Horse's-Head Nebula. The bright over-exposed star is Zeta Orionis or Alnitak, one of the three members of Orion's Belt; it is 1600 light-years away, and is about 40,000 times more luminous than the Sun. Photograph by David Malin, AAT

The Nebula is just visible with the naked eye in the Hunter's Sword, not far from the three bright stars of the Belt. It is close to the equator of the sky, so that it can be seen from every inhabited continent (Hawaii itself is at latitude 20°N.). It is known officially as M42, because it was the 42nd object in a famous catalogue of clusters and nebulæ drawn up by the French astronomer Charles Messier 200 years ago, and because of its beauty it is a favourite with amateur astronomers. Almost any telescope will show that the gas-and-dust cloud is wispy, and on its near side are four stars arranged in a trapezium pattern. It is these stars that act upon the nebular material, illuminating it and also making it emit a certain amount of light on its own account.

The Orion Nebula is very large, and since it is 'only' a little over 1000 light-years away from us it is particularly prominent. Its main interest is that it is a stellar nursery. Inside it, fresh stars are condensing; the process is not rapid, but even since regular observations of the Nebula began, less than a century ago, several stars have been seen to 'ignite'.

Deep inside the Nebula is the mysterious body which is known as the Becklin–Neugebauer or BN Object, after its two discoverers. It is absolutely invisible – the light from it is blocked out by the intervening

The Lagoon Nebula, Messier 8 in Sagittarius. This is an emission nebula; the material is excited to self-luminosity by the presence inside it of an extremely hot star known as 9 Sagittarii. Also seen in the nebula are dark patches known as Bok Globules in honour of Bart J. Bok, who drew attention to them in 1946. Each is about a hundredth of a light-year in diameter; they may be made up of material that is gradually collecting to form a new star, in which case they may be regarded as protostars. They are by no means peculiar to the Lagoon Nebula. This photograph, probably the best ever obtained, was taken by David Malin with the AAT.

material. However, infra-red radiation is not so affected, and can slice its way through the intervening gas and dust. Otherwise, we would know nothing about the existence of BN.

What precisely is it? Opinions differ, but on the majority view, it is a very young star that has only just started to shine. It is cool by stellar standards, but quite able to heat the dust-cloud that surrounds it. The heated dust emits infra-red, and the radiation comes through to Earth, where it is duly focused into the UKIRT detectors.

It is not unique. Other infra-red sources exist in M42, and in other nebulæ too; they are, presumably, young stars of the same type – and in fact BN provides only about ten per cent of the total luminosity of the region. The UKIRT can examine objects in various stages of extreme youth, and can prove that we really are peering into a stellar birthplace. Among other birthplaces are the beautiful Omega Nebula in the constellation of Sagittarius (the Archer) and the equally lovely Trifid Nebula, also in Sagittarius.

This may be the appropriate moment to say more about stellar evolution in general, because it is another subject about which our views have been considerably modified during the past 25 years. And let it be made clear at once that the whole life-story of a star depends on its initial mass. There are four main classes: (1) stars much less massive than the Sun. (2) stars about equal to the Sun; (3) stars several times more massive than the Sun; (4) even more massive stars, which have relatively short, violent careers.

The Hertzsprung-Russell or H-R Diagram. The horizontal scale shows the spectral type and surface temperature; the vertical scale shows (*left*) the luminosity of the star compared with that of the Sun and (*right*) the absolute magnitude of the star. Absolute magnitude is defined as being the apparent magnitude that a star would appear if it were seen from a standard distance of 32.6 light-years. Thus the absolute magnitude of the Sun is about + 5; ZAMS or zero-age main sequence show the positions on the diagram where young stars settle to their long period of stable hydrogen-burning. Important parts of the diagram are shown together with the positions of several well-known stars.

Stars of Class 1, between 1/16 and 1/100 the mass of the Sun, condense gravitationally in the usual way, but never become hot enough for nuclear reactions to begin inside them. The critical temperature at the core is of the order of 10,000,000°C., and if this is not reached the star will merely glow feebly before fading away to end as a cold, dead globe.

With stars comparable with the Sun, things are different. Gravitational forces lead to condensation; the star becomes hot inside, and after a period of instability the core temperature becomes high enough for nuclear reactions to be triggered off. As I have already said, the essential 'fuel' is hydrogen, which is turned into helium with resulting release of energy and loss of mass. (This is usually called hydrogen 'burning', though in some ways the term is misleading.) The star settles down to a long period of steady, stable existence, and joins what we call the Main Sequence.

The Main Sequence is a feature of the H-R or Hertzsprung-Russell Diagram, devised by Ejnar Hertzsprung of Denmark and Henry Norris Russell of the United States in the early years of this century. In an H-R Diagram, stars are plotted according to their luminosities and their spectral types. There are several classes of stellar spectra, each denoted by a letter of the alphabet – O, B, A, F, G, K, M – but the important point is that we have a means of classifying stars according to their surface temperatures, O and B stars being the hottest and M the coolest. The

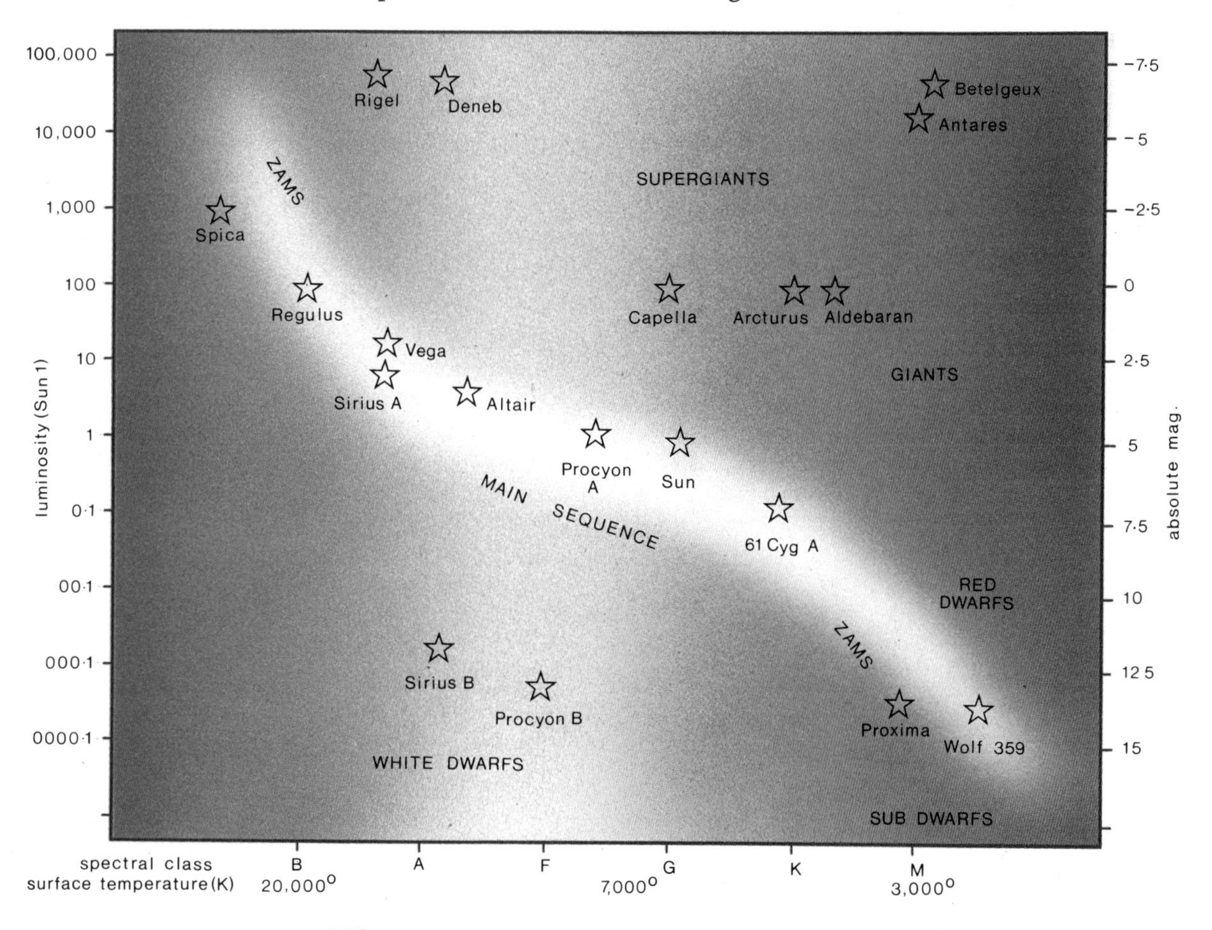

colours of the stars also depend on their surface temperatures: O, B and A stars are white or bluish, F and G yellow, K orange, and M red. The Sun is a typical Main Sequence star of Type G.

Plentiful though it is, the hydrogen available for 'burning' cannot last for ever. After the star has been on the Main Sequence for around 10,000 million years the hydrogen runs low, and energy production stops. The helium-rich core shrinks under the influence of gravity, and heats up once more. There is still a great deal of hydrogen left outside the star's core, and the heating-up eventually causes it to start reacting again, so that we have a star with an inert helium core surrounded by a hydrogen-burning shell.

The star begins to expand, but the steady squeezing of the core raises the temperature so much that the helium starts to react. Just as hydrogen nuclei combined to make helium, so the helium nuclei now combine to make heavier elements – carbon and oxygen. Energy pours out; the star swells still further, and the surface cools down, until the temperature has fallen to about 3000°C. and the star has evolved into what is termed a Red Giant. When this happens to the Sun, there can be no hope of survival for the Earth or any of the inner planets. There are plenty of Red Giants in the sky; Aldebaran in Taurus (the Bull) and Arcturus in Boötes (the Herdsman) are typical examples.

When all the available helium in the core is used up, there is a fresh period of shrinking, and the temperature rises once more. The core, now made up of carbon and oxygen, remains inert, but the helium around it begins to react, and we have a decidedly complicated situation; two 'burning' shells, the outer one of hydrogen and the inner of helium, around the non-reacting core. The star starts to pulsate, and finally the outer layers are thrown off completely, so that we are left with the old core and a ring of tenuous gas; this is what is termed a planetary nebula, though it is not truly a nebula and it is certainly not a planet. Many planetary nebulæ are known, and they are beautiful objects, but each marks the death-throes of a star, and after a few thousands of years the thrown-off gas ceases to shine. We then have only the star's core, which has become very small and very dense, with its constituent atoms crammed so close together that the density may reach 100,000 times that of water. Stars of this kind are known as White Dwarfs. They are remarkable objects; a cupful of White Dwarf material would weigh many tons.

White Dwarfs are very common. The most famous of them, though by no means the most extreme, is the faint companion of Sirius, the brightest star in the sky. Sirius, still on the Main Sequence, has an A-type spectrum and is 26 times as luminous as the Sun; its companion is only 1/10,000 as bright, but the difference in mass is not nearly so great as might be expected – it is rather like balancing a lead pellet against a cream puff. In the fullness of time Sirius itself will become a White Dwarf, and the dying pair will continue to shine feebly for thousands of millions of years before all their light and heat leaves them, so that they become dead, invisible Black Dwarfs.

Come now to Class 3, where the initial mass is several times greater than that of the Sun. Everything happens much more quickly. We have the same sort of sequence at first, though at an accelerated rate – the

The beautiful planetary nebula NGC 6781 in Aquila, photographed with the Palomar 48-inch Schmidt telescope. Many foreground stars are shown.

preliminary shrinking and instability, the sudden ignition, the hydrogen burning and then the helium burning. But after this the story is different. The immense pressure at the core forces the temperature up to 700 million °C., and this is enough to make the oxygen there start to react; at 1000 million °C. carbon joins in, to build up nuclei of a still heavier element, silicon. At 3000 million °C. silicon reacts to produce nuclei of iron, so that we now have an iron core surrounded by successive shells of silicon, oxygen, carbon, helium and finally hydrogen (I am sorry if this all sounds rather complicated!).

Events are moving towards a climax, because iron will never 'burn', no matter how high the temperature becomes. There comes a moment when the core can no longer withstand the immense pressures which are crushing it. The iron atoms are broken up, and the result is disaster.

An ordinary atom may be likened to a miniature Solar System; there is a central nucleus containing particles called protons, each of which carries a positive charge of electricity, around which orbit less massive particles called electrons, each of which has a negative charge. In a complete atom there are just enough negatively charged electrons to balance the positive charge in the nucleus, so that the atom is electrically neutral. (I am well aware that this description is over-simplified: neither a proton nor an electron should be pictured as a solid lump, and atomic nuclei can be highly complicated affairs, but my description will do for the moment.) In the collapsing star we have only a kind of chaotic ocean of broken-away protons and electrons, and when the pressure is sufficiently great these particles are forced together, combining to make up neutrons. A neutron has no electrical charge: $+1-1 = 0$.

When this point is reached there is a cataclysmic implosion (the opposite of an explosion), and further energy is released by a flood of the strange, almost or quite massless neutrinos. The shock-wave set up in the globe literally blows the star to pieces, and for a while it shines with millions of times the power of the Sun. It has become a supernova, and as the outburst dies down – which takes several months or even years – all that remains is a cloud of expanding gas, in the midst of which is the neutron star which represents the remnant of the originally huge and powerful sun. The density of the neutron star is so great that a cupful of it would weigh at least 100,000 million tons.

It is difficult to picture what a neutron star can be like. Possibly the outer layer is rigid, and made up chiefly of iron; inside is the neutron-rich material, and in the centre a core of unknown composition. It has been said that a neutron star is like an elderly raw egg inasmuch as it has a solid shell and several peculiar fluids inside. Remember too that it is very small, with a diameter of only a few miles.

Supernovæ are seen now and then, though not all are of the same type. In our own Galaxy four have been recorded during the past thousand years; those of 1006, 1054, 1572 and 1604. The 1006 star was the most brilliant of them, and at its peak seems to have been comparable in brightness with the quarter-moon, but records of it are very scanty, mainly because it flared up in the southern hemisphere of the sky (in the constellation of Lupus, the Wolf) and most of the contemporary star-gazers were in the north. The 1054 supernova has left a patch of gas known to all astronomers as the Crab Nebula; it lies in Taurus (the Bull). A small telescope will show it, though photographs taken with large instruments are needed to bring out its complex structure. It is over 6000 light-years away, so that the actual outburst happened more than 6000 years before anyone on Earth saw it. Both the two later supernovæ were spectacular, and became bright enough to be visible with the naked eye in broad daylight, though they have left no Crab-type patches and apparently no neutron stars. It is a pity that no supernovæ in the Milky Way system have appeared since 1604; astronomers would dearly like to study one with modern equipment, though preferably from a respectful distance! A new galactic supernova may appear at any moment, and by the law of averages one seems to be overdue. Meanwhile, we have to be content with observing supernovæ in other galaxies, many millions of light-years away.

The Vela supernova remnant. The photograph shows a portion of a roughly circular shell of fine luminous filaments in the southern constellation of Vela. The filaments outline the present position of the still-spreading blast wave from the detonation of a supernova some 10,000 years ago. Passage of the shock wave heats the tenuous interstellar gas, causing it to emit visible light. Offset from the centre of the shell is the pulsar 0833-45, a rapidly-spinning neutron star, only a few miles across which is believed to be the remains of the exploded star itself. The Vela Pulsar is only the second pulsar to have been detected optically, although some hundreds have been found with radio telescopes. Photograph by David Malin, AAT

The Crab Nebula is one of the most informative objects in the sky. It sends out radiations at all wavelengths, from the very short to the very long, and it is strong in the infra-red. Its 'power-house', the central neutron star, was first detected optically by astronomers at the Steward Observatory in Arizona in 1969, and appears as an extremely faint speck,

flashing 30 times a second because it is spinning rapidly. It had already been tracked down because of its radio emissions, and was the first optically-identified 'pulsar'.

I will have more to say about pulsars shortly. Meanwhile, what about the extremely massive stars, those of Class 4?

Here the situation is different again. Once the final collapse starts, nothing can halt it, and the star does not even have the chance to explode as a supernova; it merely goes on shrinking and shrinking, becoming denser and denser – and as it does so, the escape velocity increases.

Escape velocity, you will recall, is the speed required for a body to break free from a planet, a star or any other object. With the Earth, escape velocity is 7 miles per second; with the Sun, rather over 380 miles per second, and so on. The smaller and more massive the body, the higher its escape velocity. With the collapsing giant star, there comes a stage when the escape velocity has reached 186,000 miles per second. This, of course, is the velocity of light; and if light cannot escape from the old star, then certainly nothing else can, because light is the fastest thing in the universe. The star has surrounded itself by a region which is to all intents and purposes cut off. It has become a Black Hole.

Black Holes have been much in the news over the past ten years or so, and there have been all sorts of curious theories about them; for instance, it has been suggested that if one could go inside one, it might be possible to emerge either into a different universe or into a different part of our own universe. Frankly, I am not impressed, quite apart from the fact that plunging into a Black Hole would be a somewhat risky business! We cannot even be sure that Black Holes really exist, though most astronomers believe that they do. They cannot be studied directly, because they cannot be 'seen' at any wavelength, so that our only hope of detecting them is by their effects on nearby bodies that can be observed.

Up to now, the best candidate is a peculiar system known as Cygnus X-1. The X indicates that it is a source of X-rays, and to take the story further we must leave the Earth and go into space.

# 13 Into the Short Waves

I very much doubt whether Cygnus X-1 would excite you if you saw it. I looked at it recently through my own telescope, and it gave every impression of being an innocent star with absolutely nothing to single it out from the other stars nearby. And yet it is of special interest inasmuch as it may well be accompanied by a Black Hole, and information about it has been collected at the Einstein Observatory.

So far, during the course of this book, I have been describing observatories which I know personally. Flagstaff, Palomar, Hawaii and the rest are no strangers to me, but I have never visited the Einstein Observatory, and it would be rather difficult for me to do so, because it is only 21 feet long, and is orbiting the Earth at a height of about 500 miles above the ground! We do not even know exactly where it is now; it was launched on 13 November 1978, and its transmitters finally failed in the spring of 1981, so that it has joined the vast crowd of dead orbiters in outer space.

Einstein was concerned with the X-ray emissions that come from the sky. They had been known ever since 1948, when X-radiation was detected from the Sun by instruments carried aloft in balloons, but it was only in 1963 that X-ray astronomy proper started. This is because the incoming X-rays are effectively blocked by the Earth's atmosphere, and in particular by water-vapour. Even going to the top of Mauna Kea will not help, and so the only course is to use space vehicles. Rockets were first used, and confirmed that there are some powerful sources in the sky, but satellites were clearly much more effective, because they could orbit the Earth for weeks, months or years instead of crashing back to the ground within a few minutes. The first X-ray satellite, Uhuru, was dispatched in 1970, and located various sources of X-radiation, among them Cygnus X-1.

The visible star is known by its catalogue number, HDE 226868. It had not previously attracted any attention, and was not expected to send out X-rays – but it did, and the Uhuru observations were conclusive.

An X-ray telescope makes use of what is termed 'grazing reflection', in which the incoming radiations are, so to speak, skimmed off the surface at a low angle, much as a flat stone will do when bounced across calm water in the manœuvre known popularly as ducks and drakes. After being collected the X-rays pass down a gold-plated tube into special detectors, where they can be analyzed and the results transmitted to the ground. Uhuru made use of two X-ray telescopes, and HDE 226868 proved to be very unusual indeed.

The first picture of an X-ray source, Cygnus X-1, obtained from the High Energy Astronomy Observatory satellite (HEAO-2) on 18 November 1978, only five days after the satellite had been launched. This is a computerized representation, which is why the stars appear as squares.

It was already known that the visible star was very remote, around 6500 light-years from us, and very luminous. It was also known to be a member of a two-body or binary system. Visual binaries, or pairs of stars, are common enough (Mizar in the Great Bear's tail is a good example), but HDE 226868 was different, because the secondary star or companion could not be seen at all, and could be located only by its effects on its shining primary. From the mutual shifts of the two, it was established that the visible star has a mass 30 times that of the Sun, which is very high by stellar standards, and that it has a diameter of over 10,000,000 miles. The invisible secondary was found to be 15 times more massive than the Sun, so that it ought to have been visible. Unquestionably we were dealing with an exceptional sort of system.

The X-radiation was markedly variable, and even seemed to flicker in a curious manner. Uhuru studied it closely, and further studies were carried out by later X-ray satellites, notably the British Ariel vehicles and, more recently, Einstein. Gradually astronomers began to build up a coherent picture, which has received general approval even though we cannot yet be certain that it is correct.

If we are right, then the companion of HDE 226868 is a Black Hole – that is to say, a highly evolved star which has passed through its Main Sequence and giant stages, finally collapsing without suffering a supernova outburst. It retains its original mass, and therefore its pull of gravity, but not even light can escape from it, which is why it remains so obstinately out of view.

What, then, about the X-rays? Astronomers have an answer here too. The Black Hole is tearing gas away from its visible companion, and this gas spirals towards the Black Hole, finally arriving at the 'event horizon', i.e. the point where the escape velocity becomes equal to the speed of light. The torn-off gas is making for the Black Hole, but there is so much of it that not all of it can be funnelled in at once, bearing in mind that the Black Hole is extremely small compared with a normal star. Therefore, the gas starts to swirl around the event horizon, becoming violently compressed and heated. High temperature, as we have seen, leads to X-ray emission – and this is what happens, accounting very well for the short-wave radiations picked up by Uhuru, Ariel, Einstein and their kind.

When this theory was first proposed, many astronomers were sceptical about it, and for that matter some of them still are. (I know several leading authorities who have no use at all for Black Holes, and class them in the same genre as science fiction.) But what is a reasonable alternative explanation for the strange system of Cygnus X-1? Nobody has yet come up with a good solution.

Moreover, HDE 226868 is not unique. There are several other X-ray sources of the same type, and these too may involve Black Holes, though admittedly the evidence is not so strong.

The idea of mass transfer, or the pulling-off of material from one component of a binary by the other, is well established, and has been used to explain the phenomena of novæ or 'new stars' which flare up now and then. The last bright nova was that of 1975. I went out to observe soon after dark one evening, as I usually do, and when I looked at Cygnus my attention was at once caught by a bright newcomer that had certainly not been there 24 hours earlier. Having satisfied myself that I was neither mad nor being deceived by a slow-moving artificial satellite, I went indoors and telephoned Herstmonceux Observatory. Yes, they knew about the nova; it had been discovered several hours earlier from Japan, where the night fell earlier than ours. I suppose I was about seventieth in order of priority, but at any rate the nova was there, and it remained visible with the naked eye for almost a week before fading back to obscurity. It has now become very dim indeed.

With a nova of this sort we again have a binary system, but this time involving two light-emitting stars instead of one normal star together with a Black Hole. The secondary component of a nova is one of the small, superdense White Dwarfs. It pulls material away from its swollen companion, and this material builds up in a ring round the White Dwarf until it has become so compressed, and so hot, that nuclear reactions are triggered off. There is still a certain amount of unused hydrogen left, so that we are back to the familiar hydrogen-into-helium process, and the combined light of the system flares up for a brief period before the outburst is over. The whole situation is very different from that involving a Black Hole, but the underlying principle is much the same.

Obviously we can never see a Black Hole directly, because it gives out no light or anything else, and if a Black Hole were moving through space

The British artificial satellite Ariel 5, which was an X-ray satellite and proved to be extremely successful. It was launched in 1974, and when it finally fell back to destruction in the atmosphere, on 14 March 1980, its makers and users declared that it was like losing an old friend! Its successor, Ariel 6, was sent up on 2 June 1979, not from Canaveral, but from the launching ground at Wallops Island in Virginia.

on its own, so to speak, we would be quite unable to track it unless it came near enough for its gravitational pull to be felt. For this reason, it seems hopeless to try to find out how many Black Holes there are in the Galaxy. However, there is one other possible method of identification, involving a binary system of different type.

Close to the brilliant Capella, which is almost overhead in the evening sky during winter in the Northern Hemisphere, there is an inconspicuous naked-eye star known as Epsilon Aurigæ. Its light is not constant, and every 27 years it slowly fades, remaining well below its normal brightness for several months before recovering. Clearly it is being covered up by something or other, and we are again dealing with a binary system. When the secondary member of the pair passes in front of the luminous star, the light which we receive on Earth is reduced.

Epsilon Aurigæ itself is another very powerful star, at least 60,000 times more luminous than the Sun, but the companion sends out no visible light at all. Neither does it emit X-rays, but it is a source of infrared radiation. There are two possible explanations. Perhaps the companion is a Black Hole, too far from its primary to pull off gas in sufficient quantity to yield X-rays, but with a 'forbidden zone' round it large enough to cover up its primary for a considerable time. Or, alternatively, the secondary may be a very young star, still in the process of contracting towards the Main Sequence, and not yet hot enough to shine in the visible range. On the whole the latter explanation seems to be the more likely, in which case the secondary is truly vast; if it were placed in the centre of the Solar System it would swallow up all the planets out as far as Uranus, though its density will be very low (millions of times less dense than the air we breathe). The best we can say at the moment is that Epsilon Aurigæ may be a Black Hole candidate, though the evidence is extremely uncertain.

If we are to make positive identifications of Black Holes, the X-ray method seems to be the best, but now that the Einstein Observatory is defunct there are no major X-ray satellites in orbit, and we must await the next launchings, which will undoubtedly take place before long, despite the financial cutbacks in the space-research programme.

If Black Holes really exist, they are the most bizarre objects in the entire universe. It is hopeless to try to visualize the conditions inside them; all the laws of ordinary common sense break down. For that matter, what happens to the old star once it has, so to speak, 'pulled in' space round it, and has cut itself off in its own particular ivory tower? I have said that the gravitational collapse is so complete that nothing can halt it; carried to the extreme, this would mean that the star would finally crush itself out of existence altogether. Alternatively, it may be that for various reasons a little energy can leak away, so that the Black Hole will eventually explode, leaving nothing behind. We have to admit that we are still very much in the dark, and I repeat that the very existence of Black Holes is sometimes questioned.

Science fiction writers have made great play of the theme of a Black Hole approaching the Solar System, and of which we would have no advance warning until its gravitational effects started to make themselves felt. Candidly this seems a most unlikely event, though I suppose

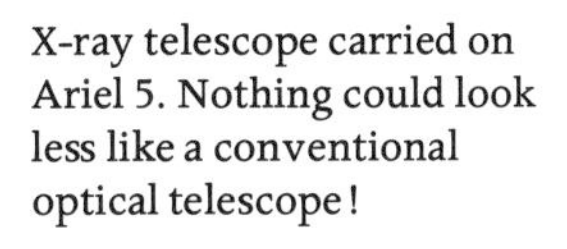

X-ray telescope carried on Ariel 5. Nothing could look less like a conventional optical telescope!

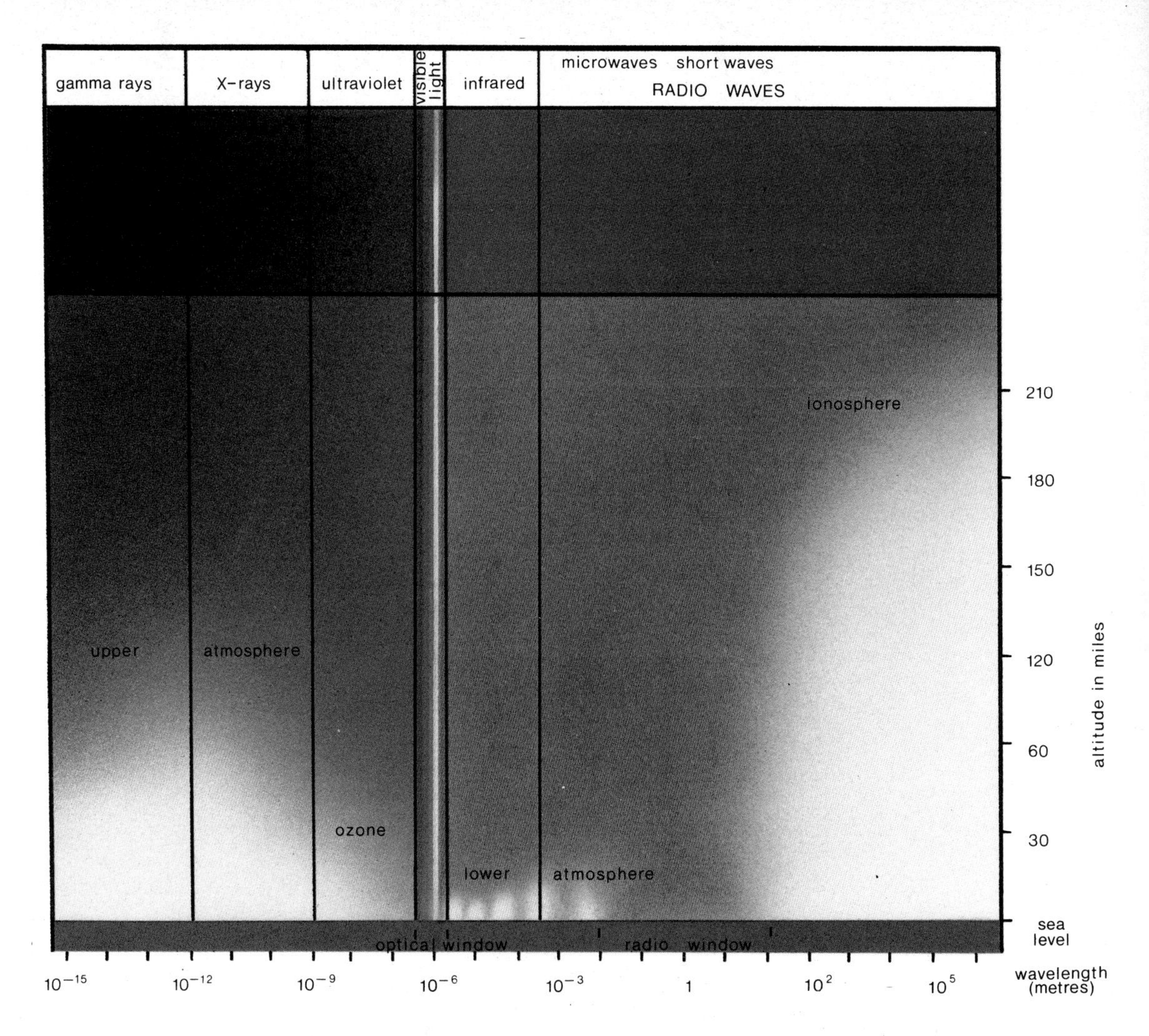


The electromagnetic spectrum. Also shown is the varying transparency of the atmosphere in different regions of the spectrum. Thus the atmosphere is transparent to visible light and some parts of the infra-red and radio regions, but long-wavelength radio waves are reflected back into space by the ionosphere, while ultra-violet is absorbed by the ozone layer, and X-rays and gamma-rays are absorbed by the upper atmosphere. Diagram by Paul Doherty.

OPPOSITE: The gamma-ray telescope at Mount Hopkins. I took this photograph in January 1980.

that if isolated Black Holes are wandering around the Galaxy we might come within range of one of them sooner or later. Meanwhile it does look as though there may be a Black Hole in the centre of the Milky Way Galaxy, because X-rays are picked up from there too. If so, the same might be true of other galaxies which are powerful emitters not only of X-radiation, but of visible light and radio waves as well.

Before leaving the ultra short-wave part of the spectrum, I must pause to say something about gamma-rays, which are even shorter than X-rays and are very difficult to 'catch'. Again we have to go above the atmosphere for most of our research, and gamma-ray satellites have been launched, the first of which went aloft as long ago as 1967. The detector this time is not a telescope of any sort, but what is termed a spark chamber, made up of electrically charged plates stacked one on top of the other. When a gamma-ray strikes the top plate it is broken up, and the particles produced cascade through the stack, producing sparks which can be photographed (hence the name).

Generally speaking, X-ray sources also produce gamma-rays, and our old friend the Crab Nebula is one of the strongest sources in the sky, but

even here only about five gamma-rays can be recorded per night, and I am reminded of the Gamma-Ray Astronomer's Hymn as given by Dr Jocelyn Bell-Burnell, discoverer of pulsars:

> Through the night of doubt and sorrow
> Onward goes the pilgrim band,
> Counting photons very slowly
> On the fingers of one hand

– a photon being a small parcel of gamma-ray energy.

There may well be an association between gamma-rays, X-rays and Black Holes, but as yet it is too early to be certain of anything, because gamma-ray astronomy is still in its extreme infancy. We do not even know where most of the gamma-rays originate; there is certainly a general background of them as well as the various discrete sources such as the Crab. The very highest-energy gamma-rays can be recorded from ground level, and equipment for this purpose has been set up at Mount Hopkins on a small piece of level surface rather below the square building of the MMT, but we must await further information before trying to come to any hard and fast conclusions.

# 14 Nebulae and Galaxies

So far as European and United States astronomers are concerned, it is unfortunate that many of the most interesting objects in the sky lie so far in the south. True, there is no bright south polar star to compare with the Polaris of the north – the nearest naked-eye candidate for the title is the obscure Sigma Octantis, which is by no means easy to identify even when you know just where to look for it – but the surrounding regions are very rich. There is, for instance, the Southern Cross, shaped more like a kite than an X, with its two brilliant Pointers, Alpha and Beta Centauri; Alpha Centauri is actually the nearest bright star beyond the Sun, and lies at a mere 4.3 light-years from us (Beta Centauri is very luminous and remote, so that again we are seeing a line-of-sight effect, and the two Pointers are not true neighbours). There is also the brilliant Canopus,

Omega Centauri, the finest of all globular clusters. It is one of the many spectacular objects in Centaurus, which is a particularly brilliant and interesting constellation. Omega lies well north of Alpha Centauri and the Southern Cross, but it never rises from anywhere in Europe.

OPPOSITE: A time-exposure showing star trails, with the dome of the AAT in the foreground. The stars shown here are circumpolar; that is to say they never set over Siding Spring, just as the Great Bear never sets over London or New York.

second only to Sirius in splendour, plus exceptionally bright regions of the Milky Way, the best examples of the spherical star systems called globular clusters, and, perhaps above all, the two Clouds of Magellan.

All these are inaccessible from Europe and almost the whole of North America. Therefore there is an obvious need to 'go south', and the emphasis during the past quarter-century has been largely on southern-hemisphere telescopes. I have already referred to those in Chile and South Africa, but to my mind the most impressive of them all is in New South Wales.

Australia has always been in the forefront of astronomical research, and major observatories have been set up, notably at Mount Stromlo, near Canberra, where the main instrument is a 150-inch reflector, completed in 1972. A new giant optical reflector was planned in the late 1960s, as a joint project between Great Britain and Australia. After long and detailed discussions, the site chosen was Siding Spring Mountain, in the Warrumbungle range of hills not many miles from the little town of Coonabarabran – made up of volcanoes which have been extinct for the past 13,000,000 years or so. Australia has no really high peaks, but Siding Spring rises to over 4000 feet, and in general the skies are clear, with steady, transparent air. Close by is the Warrumbungle National Park, a place where kangaroos, wallabies and other kinds of roos can live in peace. (Driving along the wood-bounded road up to the Observatory can be quite a hazardous business in the dark, because kangaroos have remarkably little road sense, and are always apt to

BELOW: Some of the domes at Siding Spring, with the Warrumbungle Mountains in the background. I took this picture from the catwalk of the AAT dome, from which there is a truly magnificent view.

The UK Schmidt telescope at Siding Spring; my photograph, December 1981.

bound out without warning, so that most cars are fitted with what are termed 'roo bars' in front of their bumpers.) Add gaily-coloured birds and the occasional koala bear, and you have an attractive scene. During a visit to Siding Spring in December 1981 I actually saw a koala, sitting in a tree close to the main dome and regarding me with no more than passing interest, though unfortunately it showed a marked reluctance to be photographed.

At present there are several telescopes at Siding Spring, including the UKS or United Kingdom Schmidt, an outstation of the Royal Observatory

The AAT; my photograph, December 1981. Comparisons are always odious, but I will give my view that even if the AAT has its equals it certainly has no superiors anywhere in the world.

Edinburgh, which has recently been used for a comprehensive photographic survey of the entire southern sky. There is also another Schmidt, Swedish in construction and linked with the Uppsala Observatory, plus 24-inch and 16-inch reflectors. But pride of place must go to the 153-inch Anglo-Australian Telescope or AAT, which has no peer anywhere in the world today.

In design it is orthodox enough. There is the usual skeleton tube and equatorial mounting; the telescope assembly weighs 335 tons, and has 275 tons of moving parts, while the dome, 120 feet across, weighs a full

500 tons. To guard against any possible shaking, the telescope is based separately from the building, and this is a wise precaution even in the absence of Hawaiian-type earthquakes. The main mirror is made of Cervit, mentioned earlier. It has been said that the light-grasp of the AAT is great enough to detect the flame of a candle at a distance of over 1000 miles. It can certainly record stars 10,000,000 times fainter than those that can be seen with the naked eye.

Incidentally, the AAT can undertake infra-red work as well as ordinary observation, and it is sometimes used in the day-time, though there can be unexpected hazards! Dr David Allen, who has been at the Observatory for over seven years, told me that when making day-time infrared observations he was apt to pick up the 'heat' given off by swallows flying at some distance from the dome.

Perhaps the most important point about the AAT is that it is completely computerized. As with most other modern telescopes, the observer spends his time in the adjacent control room rather than in the dome itself, but with the AAT computerization is unusually complete. Tom Cragg, the Senior Night Assistant, demonstrated to me the way in which the telescope is set. The required co-ordinates are plugged in; at the touch of a switch, the telescope swings slowly but smoothly to the position indicated, and the dome moves until the slit is in the correct place, so that the object to be studied is shown on the television screen in the control room. The setting is accurate to within two minutes of arc, and a slight manual adjustment is enough to bring the object into the centre of the field of view. Then the driving mechanism takes over, and work can begin without any danger of the object shifting out of the field.

Every possible precaution is taken to ensure complete accuracy. Computers take care of the very slight flexing of the mirror as the telescope is moved around; changes in atmospheric pressure are also compensated for, and the temperature in the dome is kept constant by the use of huge fans below the floor, which suck in air through the slit before work begins and ensure that there is no difference between the temperature inside and outside the dome. All in all, I suppose that the AAT is the most computer-controlled telescope in existence, and after a night's work the observer takes away a reel of magnetic tape rather than a notebook.

At least, this is true for most branches of research, where electronic detectors have taken over. However, the AAT is also used to take photographs, which are not only magnificent (probably the best ever obtained) but are much more than mere 'pretty pictures'.

David Malin, who came to Siding Spring expressly to carry out photographic work, has been responsible for some entirely new techniques. Generally speaking, he does not use colour film, but takes black-and-white pictures, using colour filters and then combining the results. The reproductions are faithful, though some people who see the photographs and then look direct through telescopes are apt to be disappointed, because the beautiful hues do not show up. Consider, for instance, the Orion Nebula, that gas-and-dust cloud in the Hunter's Sword which contains the mysterious BN. The pictures show it vividly coloured, but to the visual observer it looks white, because the intensity of the light is too low for the colours to become detectable.

Nebulæ are particularly lovely when photographed in this way. As we have seen, many of them are stellar birthplaces, and in some of them it is possible to make out small, dark patches which may well be stars that have not yet condensed sufficiently to start shining: proto-stars, in fact. One object which has always fascinated me is the so-called Keyhole Nebula, near an extraordinary star called Eta Carinæ. Here one really can see colour when looking direct through even a modest telescope. Eta Carinæ itself does not look like an ordinary star at all; it reminds me of a red blob.

It has a long and unusual history. It was included in a star-catalogue drawn up as long ago as 1603, when it was regarded as a perfectly normal member of the constellation of Carina (the Keel of the Ship, which also contains Canopus). Later it was found to be somewhat variable in light, but nobody was prepared for its flare-up in the early part of the 19th century, when for a while it even outshone Canopus, and was therefore the second brightest star in the sky. Subsequently it faded, and is now only on the fringe of naked-eye visibility, but the associated nebulosity is still there, together with the dark patch which really is rather like a keyhole in shape.

What precisely is Eta Carinæ? At its peak it may have shone as brilliantly as five or six million Suns put together, and it is still a very strong infra-red emitter. One possibility is that it has thrown off a dusty shell which is absorbing much of its light, so that it is still as powerful as it

The Eta Carinæ region. This is the brightest part of the Milky Way, and contains an unusually large number of young, hot stars. The nebula itself is a cloud of glowing gas, made up primarily of hydrogen, whose red emission radiation explains the colour. About 1/10 of the nebula is made up of helium, and all other elements account for a few per cent. Eta Carinæ, in the nebula, is the most luminous and probably the most massive star known. Photograph by David Malin, AAT

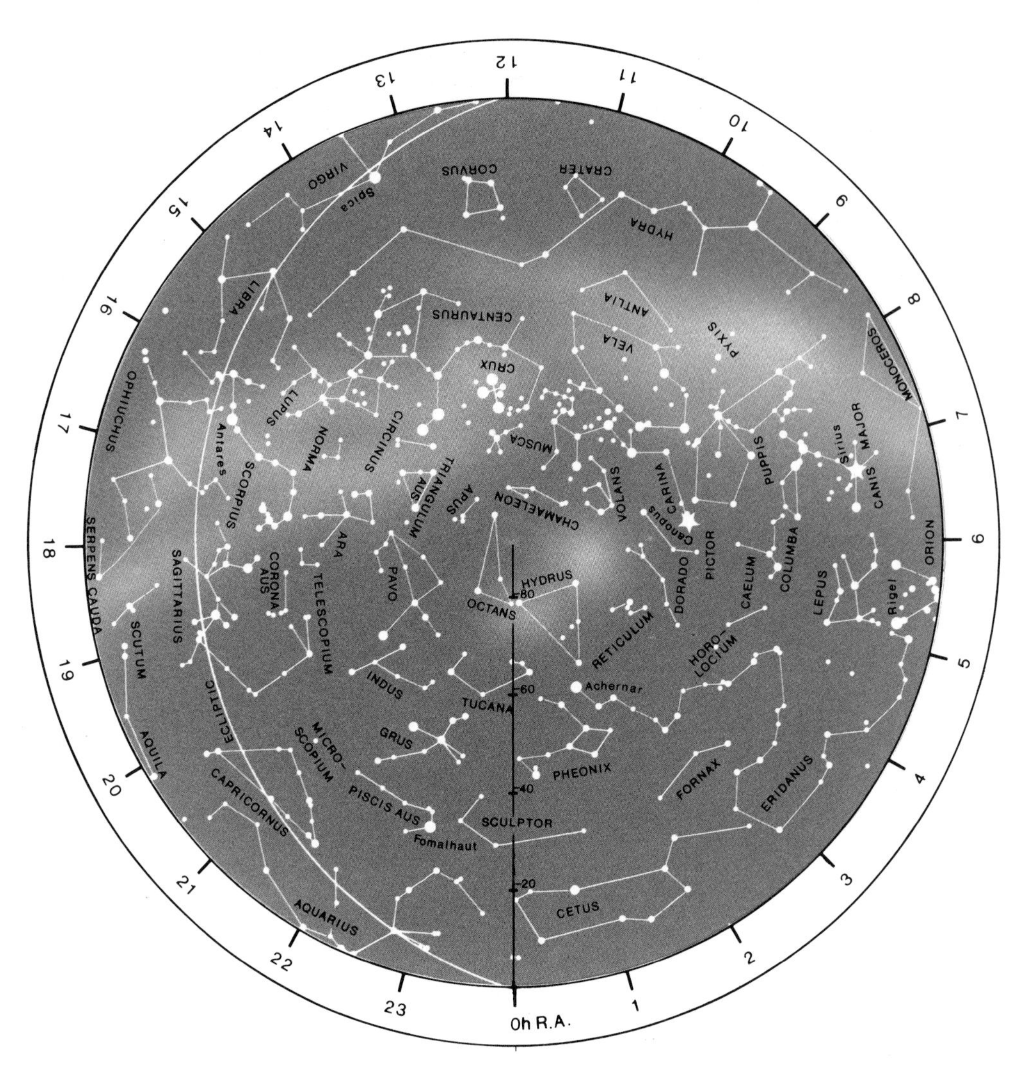


Southern sky map. The whole of the southern hemisphere is shown; the equator is marked by the outer circle. Constellation names are given, together with the names of the brightest stars. The position of the ecliptic is indicated; this may be defined as the apparent yearly path of the Sun among the stars. The figures around the equator indicate R.A. or right ascension, which may be said to be equivalent to longitude on the Earth's surface. The south celestial pole, in the middle of the circle, is not marked by any bright star. Diagram by Paul Doherty

used to be 150 years ago; alternatively it may be an exceptional type of nova, or an unusual binary system. Rather less plausibly, it has been suggested that Eta Carinæ is so unstable that it is making ready to collapse in a supernova outburst. At any rate, it is unique, and it may brighten up again at any time.

Globular clusters are also ideal subjects for spectacular photographs. The two brightest in the sky, Omega Centauri and 47 Tucanæ, are also in the far south, never to be seen from Europe but well placed for Siding Spring, and each reveals itself as a starry mass, with the individual members so crowded towards the centre that it is impossible to see them separately. In a way appearances can be deceptive, and even in the middle of a globular cluster the stars are too widely spaced to collide except very occasionally, but it is true that if our Sun lay in such a region there would be no darkness at all. The night sky would be ablaze with stars, only half a light-year away in some cases, casting shadows and even perhaps showing obvious disks. A globular such as Omega Centauri may contain upward of a million members.

Over 100 globular clusters are known in our Galaxy. They lie round the edges of the main system, and all are very remote, most of them beyond 20,000 light-years. Omega Centauri and 47 Tucanæ are prominent with the naked eye, and even in the north there is one naked-eye globular, Messier 13 Herculis, though we have to admit that it is far inferior to its southern rivals.

Not far from 47 Tucanæ in the sky, and also fairly close to the south pole, lie the two Clouds of Magellan. These are not members of our Galaxy. They are separate systems, as was first shown when the Cepheids in them 'gave away' their distances. They are the closest of the large independent galaxies, and are irregular in form; they contain objects of all kinds, including giant and dwarf stars, gaseous nebulæ and globular clusters. One star in the Large Cloud, S Doradûs, is a million times more powerful than the Sun, though it is too remote to be seen with the naked eye. Apart from Eta Carinæ, it is probably the most luminous star we know.

Recent work at Siding Spring, due largely to David Malin's improved techniques, has led to an important development in our understanding of external systems. But before coming on to this, I must say something more about galaxies in general.

Our Galaxy, with its 100,000 million stars, is a flattened system. The Sun lies about 33,000 light-years from the centre, and is moving round it, taking about 225 million years to complete a full circuit – a 'cosmic year', if you like, so that one cosmic year ago the most advanced life-forms on Earth were amphibians, and even the dinosaurs had yet to make their entry.

Seen from 'above' or 'below', the Galaxy would look like a spiral, resembling a Catherine-wheel. Many other galaxies, too, are spiral in form, including the famous Andromeda Galaxy in the northern sky, which is rather more than 2,000,000 light-years away, and is visible with the naked eye as a faint misty smudge. The Andromeda Galaxy is larger than ours, and is the senior member of what is called the Local Group, which also includes the Clouds of Magellan, a spiral in the constellation of the Triangle, more than two dozen dwarf systems, and probably another large system called Maffei 1, about which we know very little because the light from it is blocked by intervening dust spread along the plane of the Milky Way. (*En passant*, the spiral forms of some of the galaxies were first seen by the third Earl of Rosse, who built a curious but highly effective 72-inch reflector in 1845. The Rosse reflector, set up at Birr Castle in central Ireland, has long since been dismantled, but in its heyday it was much the most powerful telescope in the world.)

Not all galaxies are spiral. Others may be spherical, elliptical or irregular; there are also the peculiar barred spirals, from which the arms issue from the ends of a bar through the main plane. There is endless variety, and no two galaxies are exactly alike. Of course we can see details only in the comparatively close ones, within a few millions of tens of millions of light-years; more remote galaxies appear only as tiny patches, with no individual stars on view, and when we look out to thousands of millions of light-years it is hard to distinguish a galaxy from a foreground star.

Photographs taken with wide fields of view can show up to 1,000,000 images on one plate, and to analyse the results is a truly Herculean task. Without computer techniques it would probably be hopeless, but new measuring devices have been produced, and the information is extracted and made available. The 'Cosmos' machine, developed in Britain, can cope with a million images in a single day, whereas the same amount of work would take a human operator six years at least. There is also the Starlink network, centred at the Royal Observatory Edinburgh, in which the results obtained from different observatories (including, of course, Siding Spring) can be combined. But how are the galaxies distributed in space, and how do they move?

Again I must delve back briefly into past history. The spectrum of a galaxy consists of the jumbled-up spectra of its member stars, but the main dark lines against the rainbow background can be made out, and these are subject to the Doppler Effect. This, as you will remember, is the apparent reddening of light when the source is receding, while with an approaching source the light is 'too blue'. The actual colour-changes are slight, but the Doppler Effect is shown by the shifting of the dark lines in the spectra.

Using the 24-inch Lowell refractor at Flagstaff, from 1912 onwards, Vesto Slipher found – to his surprise – that apart from the Andromeda Spiral and a few others, now known to be members of our Local Group, all the galaxies showed red shifts in their spectra, indicating that they were moving away from us. After Edwin Hubble had used the 100-inch Mount Wilson reflector to show that the galaxies really were external systems, another curious fact emerged; the further out a galaxy lay, the faster it was receding. The invariable rule was 'the further, the faster', from which it followed that the entire universe must be expanding.

There was nothing wrong with the observations, and the Doppler Effect seemed to be the only rational explanation for the red shifts in the spectral lines of the galaxies. A definite law emerged; the recessional velocities increased with distance according to a ratio which is still called Hubble's Constant, and the idea of an expanding universe became so firmly established that even today it is regarded as heretical to question it.

Assuming that the red shifts really are cosmological – that is to say, due to the Doppler Effect – we can form some idea of the distances involved. They are staggering by any standards. Galaxies are known which are over 10,000 million light-years from us, and since the Earth is less than 5000 million years old we are seeing these galaxies as they used to be before our world came into existence. The strange objects known as quasars, or QSOs (Quasi-Stellar Objects) are generally believed to be even more remote in some cases. Actually, it now seems likely that QSOs are special types of galaxies, but since their story is so closely linked with that of radio astronomy I propose to defer it until the next chapter.

Distance-measurements can be checked to some extent with galaxies close enough to reveal their individual stars. There are Cepheids in them, and because the Cepheids are so luminous they can be seen out to millions of light-years. At distances when the Cepheids are lost to view, we can use supergiant stars, on the assumption that supergiants in any external system are likely to be of about the same luminosity as those in our Galaxy. Further away still we can watch for supernovæ, which occur quite often. Since a supernova reaches a peak power of over thousands of millions of Suns combined, it provides us with yet another useful means of distance-estimating.

All these methods confirm the results of the Doppler shifts. Yet, when we come to galaxies so remote that no stars at all can be made out, all we can do is to measure the red shifts of the spectral lines, find the recessional velocity and then use Hubble's relationship to estimate how far away they are; and this is one potential source of uncertainty, because we have no firm guarantee that Hubble's Constant is valid right out to the edge of the observable universe.

Galaxies tend to occur in groups or clusters. Our own Local Group, with its membership of around 30, is comparatively sparse, while others are much more populous; there is for instance the Virgo cluster, at about 60,000,000 light-years, in which there are thousands of members. One disquieting fact has become evident lately. There are two ways of finding the total mass of a cluster of galaxies: first by adding up the estimated individual masses of all the known members, and secondly by seeing how they move in relation to each other. At present these two methods

The spiral galaxy NGC 2997. The disk of the galaxy is inclined at about 45° to our line of sight, revealing its internal structure and making the galaxy appear oval. The two spiral arms, which seem to originate in the yellow nucleus, are speckled with bright red blobs of ionized hydrogen which are similar to regions in our Galaxy in which stars are being formed. These produce the hot blue stars which are responsible for most of the light coming from the spiral arms. Photograph by David Malin, AAT

Star-clouds and dust in the Milky Way towards the direction of the galactic centre. Photograph by David L. Talent, Cerro Tololo Inter-American Observatory, Chile

give very different results. The second method gives a much higher value than the first, so that there seems to be a great deal of 'invisible mass', and we simply do not know where it can be. There are even suggestions that it may be locked up in Black Holes . . .

Another mystery comes from observations made in 1981 by astronomers using the Kitt Peak reflector, the Palomar 200-inch, and the MMT at Mount Hopkins. During studies of the distribution of galaxies, a vast empty 'hole' was found. Over an area of space about 300 million light-years in diameter, in the general direction of the constellation of Boötes, no galaxies of any type could be found. If the observations are reliable – and as yet they have not been fully confirmed – we must suppose either that the galaxies in that region are too faint to be seen, or else that they are genuinely absent; obscuring material can hardly be the answer, because there are plenty of visible galaxies on the far side of the 'hole'. Up till now there is no explanation, and the discovery does stress, yet again, that we still have a great deal to learn about the layout of the universe.

The reason why so many galaxies are spiral in form has caused endless discussion. The spirals are less massive than the giant ellipticals, and contain greater numbers of young stars; they also contain large amounts of interstellar gas and dust, whereas in the ellipticals most of the star-forming material has been used up. Recent theories indicate that the spiral arms may be due to 'pressure waves' sweeping round the centres of the systems, so that when a pressure wave reaches any particular region the interstellar material is compressed and heated, and star formation is triggered off. If so, then our Sun itself was born in a pressure wave around 5000 million years ago.

This brings me back to Siding Spring, and David Malin's remarkable photographs.

With his new techniques, he has been able to take pictures of elliptical galaxies that show excessively faint surrounding 'shells', about 100 times fainter than the mean brightness of the night sky itself. Apparently these shells are made up of stars, and it may well be that they were produced by the collisions of two or more spiral galaxies, which merged together, lost their original shapes, and combined to make up a single, more massive elliptical system. David Allen has described them as 'fossil structures', dating back a very long way indeed. If so, then the shells are made up of stars that were thrown out during the long-continued collision.

This is a new idea, and it would certainly be premature to suggest that all ellipticals are made up of former spirals; on the other hand we do not yet know much about the evolutionary sequences of the galaxies, though it has always been doubted whether a spiral could turn into an elliptical or vice versa. The discovery of the encircling shells may turn out to be a valuable clue.

Of even greater importance, perhaps, are those galaxies that show traces of tremendous activity going on inside them. There are plenty of these, but optical astronomy cannot give us a complete picture. We must turn to radio methods, and to Great Britain.

# 15 The Radio Sky

It is not easy nowadays to remember how limited the astronomy of 1957 really was. Researchers were working in the equivalent of blinkers, because they had to deal mainly with the light that we can see. What we now call invisible astronomy was very much in its infancy.

The real breakthrough in the long wavelengths was due to one man: Professor Sir Bernard Lovell, whose grandiose schemes proposed in the late 1940s and early 1950s sounded more like science fiction than anything else, but which led to a complete change in our ideas about the universe. Without Lovell, the huge 'dish' so familiar to people living in Cheshire would not have been built. But before coming on to the remarkable story of Jodrell Bank, with its bizarre undertones (even a threat of prison!) I must once more look back briefly, though this time only to 1931.

It was then that radio astronomy began, more or less accidentally. A young American research worker named Karl Jansky was investigating 'static' on behalf of the Bell Telephone Company when he found that his improvised aerial was picking up long-wavelength signals from the Milky Way. For various reasons he never really followed up this discovery, and certainly did not appreciate the importance of it; he published a few papers, and then let the whole matter drop. Professional astronomers were profoundly uninterested, and also ignored the work of Grote Reber, the first intentional radio astronomer, who built some special equipment consisting of a 30-foot mirror and improved detectors. Operating at a wavelength of about two metres he located several discrete radio sources in the sky, but they did not come from bright stars such as Sirius or Rigel; instead they came from regions where nothing particular could be seen visually. Reber published his findings in a series of papers between 1940 and 1945, but the war was in progress, and most scientists were otherwise engaged. (I remember hearing something about the new work myself, but at that stage in my career I was thinking more about Hitler than the Milky Way.)

Radar work started at around the same time. By 1942 British anti-aircraft gunners were already using it to locate German aircraft, and it was found that some mysterious jamming was affecting the equipment. It was at first thought to come from Germany, but an investigating team headed by J. S. Hey showed otherwise; the strange signals came from the Sun, which was then near the peak of its 11-year cycle of activity. Unfortunately, Hey and his colleagues could not publish their discovery,

The Jodrell Bank 250-foot radio telescope; my photograph, November 1981. The 'dish' was pointing to the zenith.

and it was not until the end of the war that astronomy emerged from its state of suspended animation.

Before long, further radio telescopes were built, much more powerful than Reber's had been, and the results became more and more puzzling as well as more and more interesting. The Sun could be expected to be a radio source, but why did the bright stars show no long-wave emissions at all? And why did strong radio waves come from what appeared to be blank areas of the sky?

Oddly enough, the birth of Jodrell Bank – still the world's most famous radio-astronomy observatory – was associated with phenomena much nearer to home than the Sun. Meteors, which dash into the upper air and destroy themselves, produce trails which can be detected by radar, and pioneer work carried out by Professor A. C. B. Lovell, as he then was, showed that radar methods could yield important results. Helped by one of Britain's leading meteor experts, J. P. M. Prentice, Lovell set to work. The grassy field upon which the great radio telescope now stands was a good site; it was well away from artificial interference (though, alas, this is no longer true), and the early tests were encouraging. However, radar is necessarily limited, and even today Saturn remains the most remote object contacted in this way, so that the main emphasis switched to radio astronomy, and regions far beyond the Solar System. Lovell was anxious to build a really large 'dish' that would be capable of picking up signals from across distances of many light-years.

The main story may be said to have begun on 8 September 1949, when

Sir Bernard Lovell with me in the planetarium at Jodrell Bank. Pieter Morpurgo took this picture when we made a television programme with Sir Bernard a few days before he retired as Director of Jodrell Bank at the end of October 1981.

Lovell contacted H. C. Husband, head of a well-known engineering firm, and asked whether it would be possible to build a dish 250 feet in diameter, fully steerable and capable of being moved around very accurately. Other engineers had ridiculed the idea. Husband did not. His actual words were: 'Oh, I don't know. It should be easy; about the same problem as throwing a swing bridge over the Thames at Westminster.' Obviously the task would be complicated, if only because it was something completely new, but Lovell's energy and enthusiasm were irresistible. Preliminary funds were obtained, and four years later the construction began.

There are still some people who believe that a radio telescope produces a visible picture. Of course it does not; the long-wave radiations cannot affect our eyes. Neither is it possible to pick up actual noise from space, because sound-waves cannot travel in a vacuum; the hissing and crackling so familiar to those who have watched scientific television programmes is created inside the equipment itself. Basically, the dish of a radio telescope collects and focuses the long waves just as an optical telescope collects and focuses visible light. The focal point is the tip of a long metal rod or 'dipole' which sticks up from the centre of the dish; the radio waves are reflected to this point, and are amplified before being passed on to an automatic pen which records them on a moving paper drum. The end product is a trace on a chart, which may not look exciting, but is amazingly informative.

One of the worst troubles in those early days was that the resolution of a radio telescope was very low. The world's largest optical telescope was the Palomar reflector with its 200-inch mirror, but a dish-type radio telescope no bigger than this would be of very little use. Size was all-important, but clearly it led to tremendous engineering problems.

There were other setbacks, too. Not unnaturally, Lovell greatly underestimated the final cost of his 250-foot paraboloid. The design had to be changed more than once, and another crisis broke when it was found that vibration problems were much more serious than had been anticipated; the cost rose and rose, and funds ran disastrously low. Even when the telescope was being built it was constantly threatened. At one point Lovell was faced with a personal bill of £500,000, and it was seriously suggested that failure to pay it might result in imprisonment! At that stage most men would have given up, but Lovell did not, and by 1957 the telescope was almost ready. It was tested for the first time on 3 February, and the movement was found to be satisfactory, but the bills remained to be paid, and it is probably true to say that scientific as well as public opinion was far from being in Lovell's favour.

What saved the situation was the launching of Sputnik 1 on 4 October 1957. Tracking the satellite was of paramount importance, and in Britain the only instrument capable of doing this sort of work was the Jodrell Bank dish.

The 250-foot radio telescope had not been designed for anything of the sort, and even when Sputnik 1 went up Lovell had no real intention of following it, but the chance was too good to be missed. On 29 October the then Prime Minister, Harold Macmillan, announced in the House that 'within the last few days our great radio telescope at Jodrell Bank has successfully tracked the Sputnik carrier rocket', and almost overnight Lovell was transformed from an irresponsible spendthrift into a hero. The immediate financial crisis was postponed. Further appeals were made in 1958 and 1959, but the future was still in doubt. The salvation came in May 1960, not from the Government, but from Lord Nuffield. The great philanthropist telephoned Lovell, asked for the amount of the deficit (around £50,000) and settled it. Lovell still remembers his sense of profound relief when he put down the receiver. Now, at last, he could settle down to some really profitable research.

There was certainly plenty to be done, if only because there were so many different kinds of radio emitters in the sky. One of the first discrete sources to be identified was the Crab Nebula, the wreck of the 1054 supernova. It was extremely strong in the radio range, and there seemed to be some inner source of power which at first was not understood. The best suggestion came from the Russian astronomer Iosif Shklovskii, who proposed that it might be due to what is termed synchrotron emission – that is to say, radiation given off by electrons spiralling about in a powerful magnetic field. But why did this happen, and what exactly was the mysterious 'power-house'? Presumably it was the remnant of the old, collapsed star, but at that time nobody really knew.

Other old supernovæ also proved to be radio emitters. One of them was the beautiful Veil Nebula in Cygnus, which takes the form of arcs of filmy gas, and is certainly all that is left of a supernova which blew itself to pieces long before the start of recorded history. However, most of the discrete sources were much more remote, and appeared to be associated with certain special kinds of galaxies.

The irregular galaxy Messier 82, in the Great Bear, was particularly interesting, because it showed obvious signs of having suffered a cataclysmic explosion deep inside it; optical observations showed that there were fast-moving streams of hydrogen travelling outwards from the centre at velocities of up to 600 miles per second. By working backwards, so to speak, it was possible to fix the time of the explosion at 1,500,000 years ago, though since Messier 82 is 10,000,000 light-years away the real date goes back 11,500,000 years. Another exceptional source, Cygnus A, proved to be much more distant – 700 million light-years – and was different again, because the main emission came from two regions to either side of the visible object, at around 100,000 light-years from its centre. And in the southern sky, there was Centaurus A, which looked as though it might be the result of a collision between two galaxies.

The Veil Nebula in Cygnus. The colours radiated by the gaseous filaments are the result of their movement through space. Ejected from a supernova more than 50,000 years ago with an initial velocity of nearly 5000 miles per second, the gas-clouds have now been slowed to about 75 miles per second by constant collisions with atoms in interstellar space. The force of these collisions ionizes the gas – that is to say, breaks up the atoms and causes the glow. Because of the steady decline in velocity due to these collisions, the nebula will cease to glow in another 25,000 years' time. The distance from Earth is about 2500 light-years. Photograph taken with the 48-inch Palomar Schmidt telescope.

The idea of colliding galaxies was taken very seriously, and was believed to explain the strong signals in the radio range. If two galaxies met head-on, the individual stars would seldom suffer direct hits, because they were too widely spread-out, but the gas and dust between them would be colliding all the time; the situation would be a little like that of two orderly crowds moving through each other in opposite directions. It all seemed very plausible, but then the mathematicians stepped in. The amount of radio energy released in this way would be much too little to explain the strong radio signals. The theory of colliding galaxies was given up, and only recently, with the work at Siding Spring, has it started to come back into favour in a modified form.

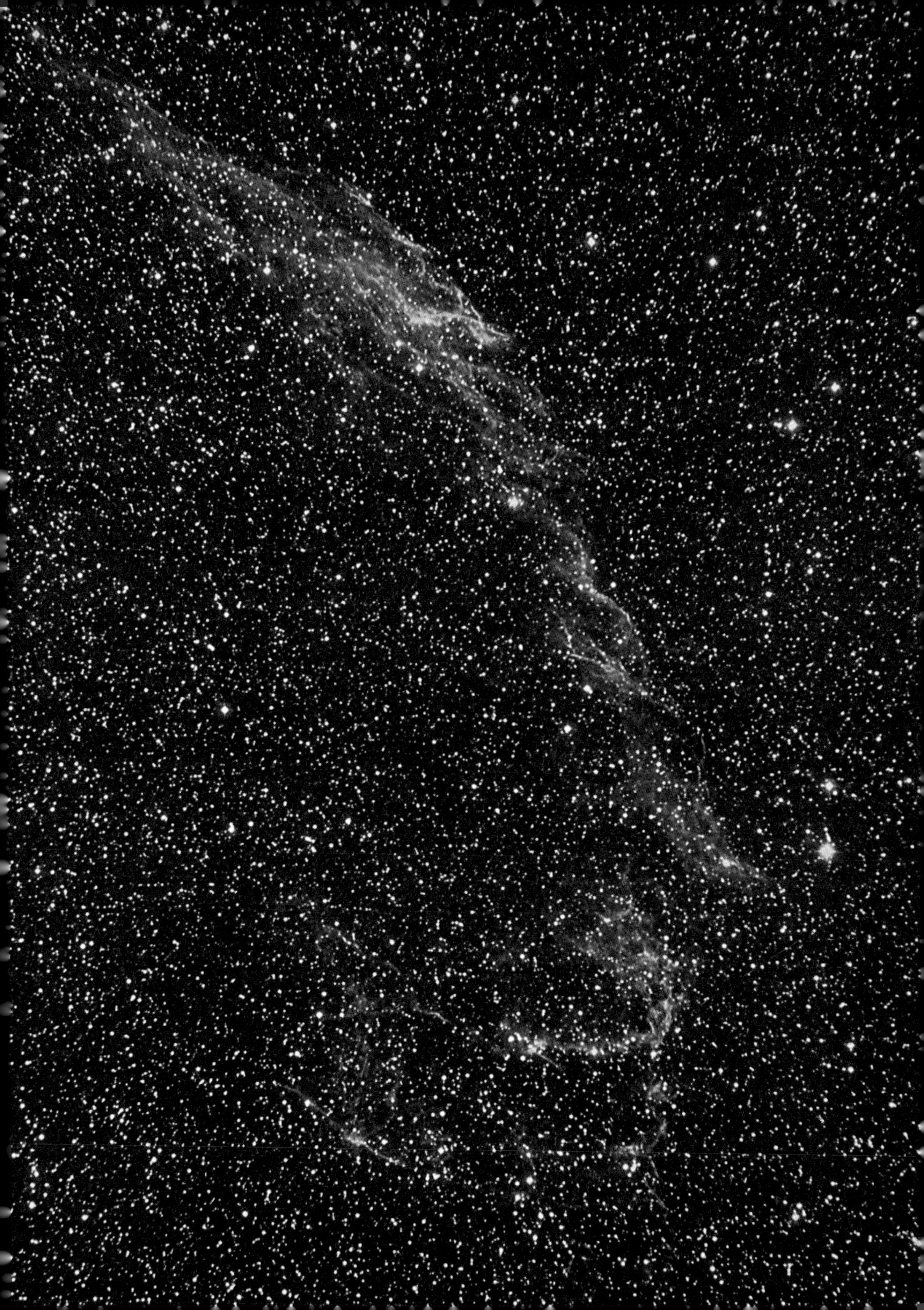

The irregular galaxy M82, in Ursa Major (the Great Bear). It was identified as a radio source by C. R. Lynds at Green Bank, West Virginia in 1961; previously the radio emission from the area had been attributed to the large adjacent spiral galaxy, M81. Using the Hale reflector at Palomar, R. Sandage showed that inside M82 there were immense, intricate structures of hydrogen gas, moving at speeds of up to 600 miles per second. It seems that a tremendous outburst occurred in M82 in the past. The distance of the galaxy is 10,000,000 light-years, and the outburst has been dated as having taken place 1,500,000 years before our modern view of it. There are few hot blue stars in M82, and it is generally held that the radio waves are due to synchrotron emission.

By the early 1960s radio astronomy was starting to find its feet. It had shown itself capable of probing out into the remote parts of the universe, and of providing information that could not possibly have been obtained in any other way. Jodrell Bank was not the only major observatory; others were set up in various parts of the world, and eventually there came the great Arecibo dish in Puerto Rico, which cannot be steered, but which has the advantage of being 1000 feet in diameter instead of only 250. (The Americans planned a fully-steerable 600-foot dish at Sugar Bowl, and actually began to build it, but gave up when it was shown that the whole engineering design was unsound.) Radio waves were picked up from the clouds of cold hydrogen spread through the Galaxy, and it was these, at a wavelength of 21.1 centimetres, which gave the final proof that the Milky Way is a spiral system. Moreover, interstellar molecules were identified, some of them organic – including alcohol!

In all these investigations Jodrell Bank, under Lovell's directorship, remained very much to the fore. A new dish was added, rather smaller than the first but very versatile, and another major step was to combine numbers of radio telescopes in what is called interferometry, so that, broadly speaking, several telescopes can be worked together, acting as though they made up a single large instrument. The main reason for this is that by its very nature, a radio telescope has less resolving power than an optical one. The longer the baseline, or distance between the individual telescopes, the greater the power of resolution. By now the baselines extend not only over limited areas, but from one continent to another.

Centaurus A (NGC 5128). This is a most unusual galaxy. The circular, uniformly bright portion includes several thousands of millions of stars, most of them yellowish and well advanced in their evolution. The galaxy is girdled by a dense dust-lane which obscures and reddens the light of stars behind it. Younger, blue stars can be seen at the edges of the dust-cloud. Centaurus A is 13,000,000 light-years away, and is a radio, X-ray and gamma-ray source as well as emitting in the visible range and in infra-red. Photograph by David Malin, AAT

Not all radio telescopes are of the dish type. Others consist of long lines of aerials, and there are even some which look rather like collections of barbers' poles. It was with one of these, not at Jodrell Bank but at the Mullard Radio Astronomy Observatory at Cambridge, that a momentous discovery was made in 1967.

The Cambridge team was led by one of Britain's most famous radio astronomers, Dr Antony Hewish. One member of his team was Miss Jocelyn Bell (now Dr Jocelyn Bell-Burnell). During the early part of 1967 a survey was being carried out when Miss Bell noticed some signals of an entirely new kind. They were very regular and very rapid, and seemed to be pulses. Little attention had been paid to them previously, but by November 1967 they had become so interesting that a special programme was run to find out more about them. The results were utterly unexpected. The pulses came from one definite point in the sky; they had a period of 1.3 seconds; each lasted for only the fraction of a second, and the 'ticking' gave the impression that they might be of artificial origin.

Radio astronomy is always beset with problems of interference, ranging from commercial transmissions to passing motor cycles, but these ticks were obviously coming from beyond the Earth, and Hewish and his team could make no sense of them. There even appeared to be a remote chance that they were deliberate signals coming from the depths of space, thereby representing our first proof of extraterrestrial intelligence. Very significantly, no announcement came from Cambridge until this idea – known in the interim as the LGM or Little Green Men theory –

The 'pulsar' radio telescope at Cambridge; it is outwardly anything but impressive, but it led to one of the most important discoveries in modern astronomy. When walking inside it, among the poles, there is a constant danger of decapitation by hidden strands of wire stretched from one pole to another. Running, or even fast walking, is emphatically not to be recommended. The grass inside the telescope is kept under control; when convenient, sheep are allowed to graze there!

had been ruled out of court. This did not take long, but at least the thought had been there, albeit only briefly.

One point soon emerged: the body sending out the signals was small by cosmical standards. Also it was definitely inside our Galaxy, and its distance was first believed to be 200 or 300 light-years, though this later proved to be an underestimate.

Jocelyn Bell's object was the first pulsar, but others were soon found, and a systematic hunt began, not only at Cambridge but also at Jodrell Bank and stations in Australia, the United States and elsewhere. The numbers of known 'ticking sources' grew rapidly, and two main theories were proposed to account for them. It seemed likely that the pulses were due to the rapid rotation of the emitting object, and the two obvious candidates were White Dwarfs and neutron stars. As time went by the White Dwarf idea was abandoned; no normal star, even a highly-evolved one, could spin quickly enough. Neutron stars, much smaller and much denser, seemed to be more promising.

The picture finally accepted, due originally to Dr Thomas Gold, was that of a neutron star which has a very powerful magnetic field, so that its radio waves can escape only by way of the two magnetic poles. As the pulsar spins, its radio beams sweep across space, rather in the manner of the beam of a lighthouse, so that every time the beam passes over the Earth we receive a pulse. The shortness of the period between successive ticks confirmed that the rotating body must be very small indeed, and this again fitted in well with the neutron star explanation. The next step was to try to identify pulsars optically, and once again attention was drawn to the Crab Nebula, where a pulsar was located in November 1968.

In the following January a serious attempt was made by astronomers at the Steward Observatory in the United States. For four nights they searched in vain, and were about to give up at the end of their allotted observing run when they had an unexpected respite; the next observer due to use the telescope was delayed, and the pulsar-hunters were able to carry on. On checking their equipment, they found that two wires had been joined up the wrong way. They put matters right, and tried again. During a brief ten-minute break in the clouds on the next night they made the vital discovery: an excessively faint object, flashing at exactly the same rate as the Crab pulsar (30 times a second) and in just the right position. The identification was complete; the Crab's long-sought power-house had been found, and soon afterwards it was also photographed from Kitt Peak.

The discovery was even more significant than might have been thought, because the Crab was known to be a supernova remnant, and it looked as though other supernova remnants might also produce pulsars. Yet this did not always seem to be true. The last two supernovæ to be seen in our Galaxy, those of 1572 and 1604, were both marked by radio sources and very faint wisps of gas, but not by pulsars. There was how-

The 'lighthouse' theory of a pulsar. Radio waves are emitted by electrons at the poles of the magnetic field of the neutron star, and leave the star in the form of a narrow beam which sweeps around as the star rotates; each time the beam sweeps across the Earth, we receive a radio pulse. Diagram by Paul Doherty

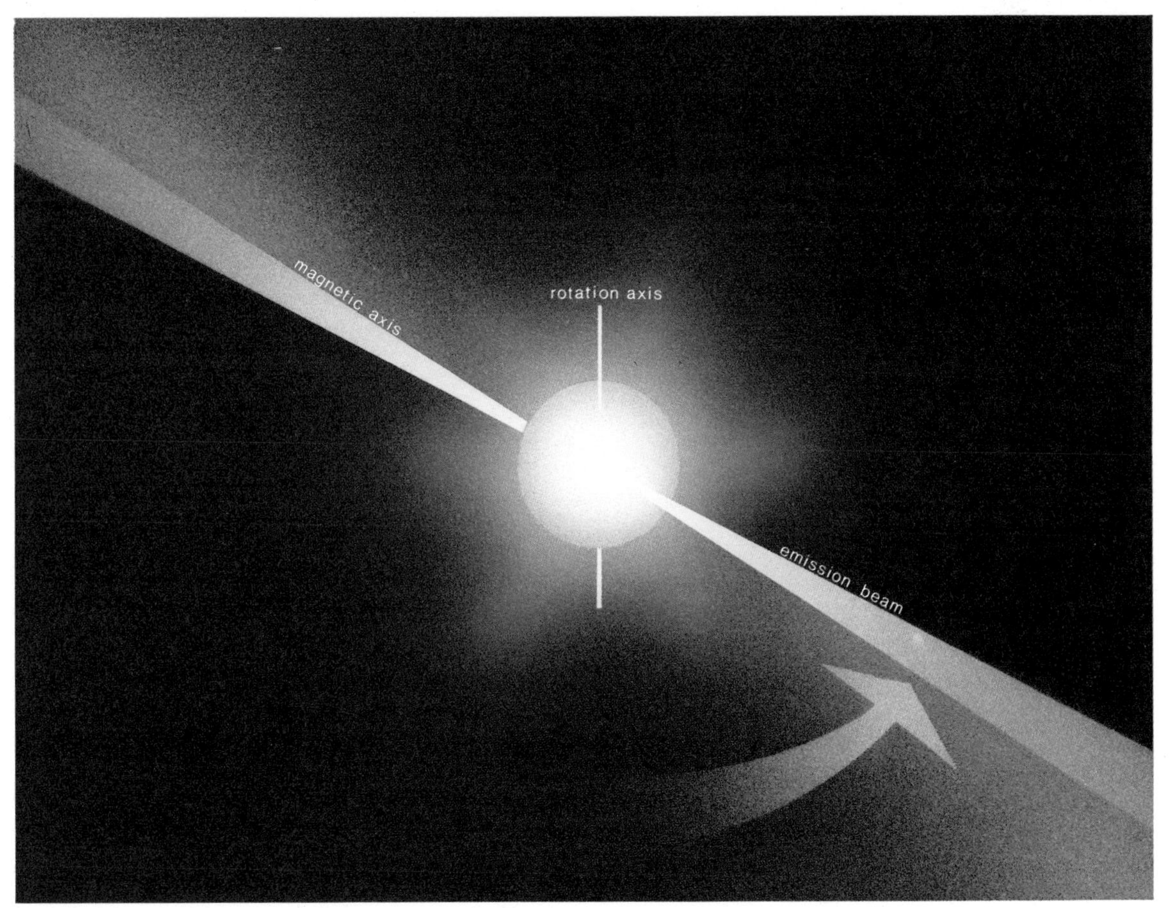

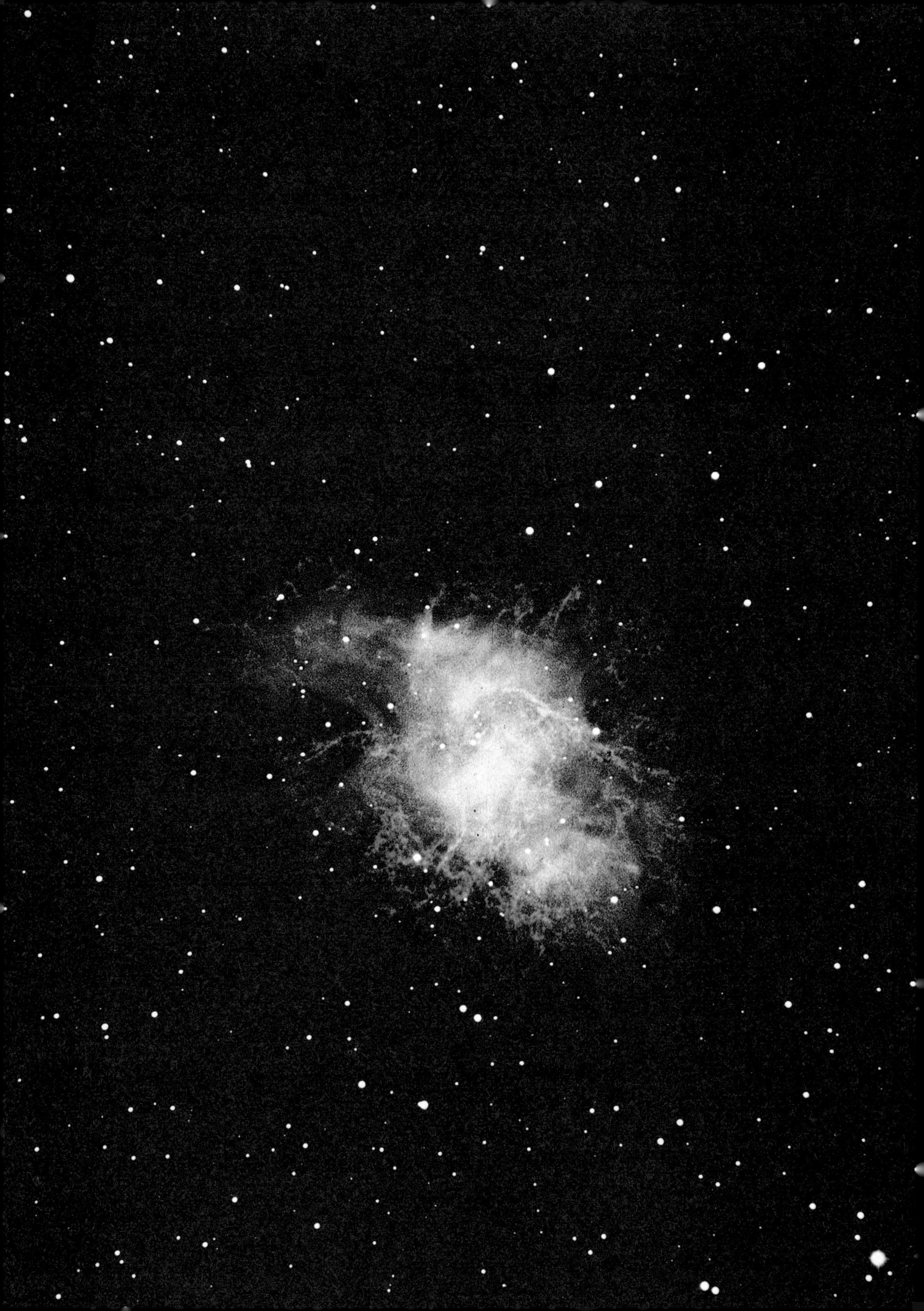

OPPOSITE: The Crab Nebula; Palomar photograph. This object, the remnant of the 1054 supernova, is perhaps the most important object in the sky from the astrophysical point of view. It sends out radiations at all wavelengths, from radio waves to gamma-rays, and it contains a pulsar.

ever another pulsar in the southern constellation of Vela, immersed in what was called the Gum Nebula – not because it was believed to be sticky, but because it had been discovered by the Australian astronomer Colin Gum. The pulse-period was 0.09 second, the third shortest known, and it was fairly clear that the nebula itself was a supernova remnant. Since the nebula is over 11,000 light-years away, the actual outburst must have happened long before there were any astronomers on Earth to observe it.

This was a problem well suited to the great light-grasp of the AAT, and it was at Siding Spring that the identification was made, in 1977. This time the pulsar was optically fainter even than that in the Crab; it was just about at the limit of detection, and in fact the Vela pulsar, as it is known, is the faintest optical object ever recorded. It is about as bright as a pocket torch seen from the distance of the Moon. It is, incidentally, a comparatively strong source of gamma-rays.

By now hundreds of pulsars have been located, though the Crab and Vela alone have been seen optically. One thing which seems definite is that as time goes by, all pulsars are slowing down, though only by tiny amounts; thus the period of the pulsar known as CP 1919 is lengthening by a thousand millionth of a second each month, so that in 300 years' time it will have grown to 1.3374 second instead of the present-day 1.3373. The inference is that pulsars lose their energy as they grow

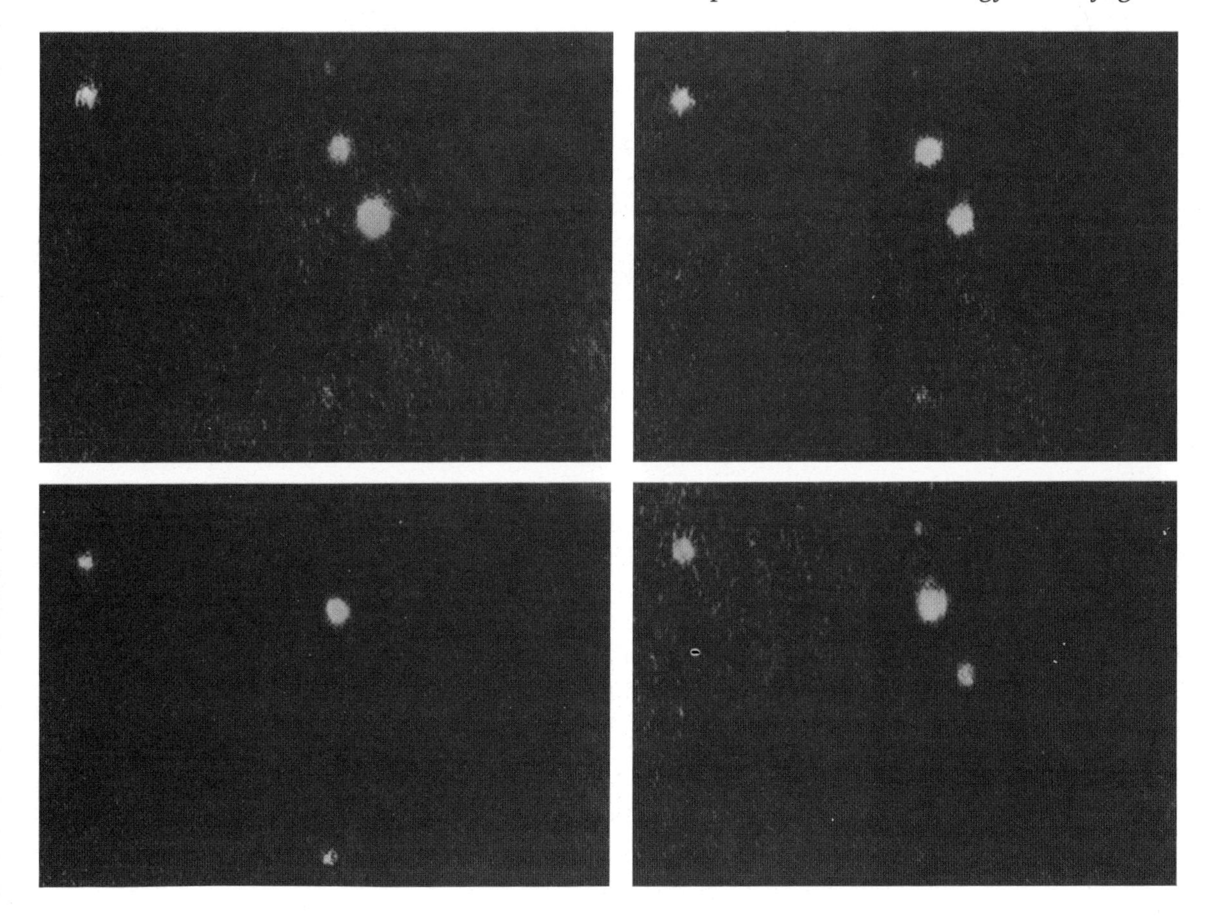

BELOW: Four different phases of the Crab pulsar; the period is 0.033 second. Photographs taken at the Lick Observatory on 3 February 1969. The pulsar is invisible in the second picture, at maximum in the fourth.

older, and rotate more slowly. The Crab pulsar, which has the shortest period known, is presumably the youngest so far found. On the other hand some pulsars show irregular changes in period; thus on 1 March 1969 the pulsar PSR 0833-45 speeded up by a full quarter of a millionth of a second. These sudden changes, termed 'glitches', are thought to be due to events in the neutron stars themselves – starquakes, if you like.

Pulsars, then, are simply neutron stars, but it is still not clear why some known supernova remnants do not produce them. It may well be that there are two types of supernovæ, and this is confirmed by optical observations of outbursts in other galaxies. Neither is it definite that all pulsars are supernova remnants, which brings me back to Siding Spring and the remarkable story of SS 433.

At Cambridge, Sir Martin Ryle (now Astronomer Royal) had led a search for radio sources near known supernova remnants, and had found quite a number. One, in the constellation of Aquila (the Eagle) had been catalogued as W 50. There was an X-ray source in the same region, and equipment carried in the British satellite Ariel had shown that the X-rays were coming from a point actually inside W 50. This seemed interesting, and it occurred to two astronomers, Dr Paul Murdin (now Project Scientist at La Palma) and Dr David Clark, that the radio source, the X-ray emitter and the supernova remnant might be one and the same. Using the AAT, they searched. On the television monitor in the control room they came across a starlike object in just the right place, and when they examined its spectrum they were certain that their inspired guess had been right. More peculiar characteristics soon showed up; the spectrum was variable, and the object turned out to be unique in present-day experience. It was then found that the star itself had been included in a list drawn up by C. Stephenson and N. Sanduleak in America, and was their 433rd entry; it was henceforth known as SS 433.

Paul Murdin and David Clark announced their results in 1979. Tremendous interest was aroused, and in fact several conferences have been held since then devoted entirely to SS 433. Murdin's picture of this weird object is of a close binary system, made up of a normal star together with what may be a neutron star but is more probably a Black Hole. The secondary pulls gas away from the primary, but not all of it can enter the Hole; some of it is rejected, so that it streams out from the Black Hole in two opposite jets, moving at a third of the speed of light. The jets wobble and spray out in a cone, after the fashion of a rotating lawn sprinkler, spinning in a period of 164 days and producing the observed changes in the spectrum. This is supported by the fact that the jets have been observed at radio wavelengths; they protrude from the central source, and rock from side to side in the same period of 164 days. The jets produce 'knots' of radio emission which move outward, again at one-third the speed of light. And to cap everything, the W 50 radio source itself is elongated, with two east-west 'ears', apparently blown by the jets like blisters on the skin of the W 50 radio shell.

If this is correct – and all the evidence points that way – then the supernova which produced W 50 did not also produce a pulsar. Instead, it produced what may be termed a 'scintar'. Yet another new type of object had come into the reckoning.

The case of W 50 confirms the growing view that supernovæ are of two different types, only one of which ends up as a pulsar. Can we therefore claim that all pulsars are supernova remnants? Again there is considerable uncertainty, mainly because there seem to be too many pulsars. A 1978 survey carried out at another Australian radio astronomy observatory (Molonglo) netted 155 in a few months. From this, it follows that there must be from one to two million pulsars in the Galaxy altogether. Remember, too, that pulsars – *as* pulsars – may be relatively short-lived; over a period which is brief by cosmic standards they will lose their energy, slow down, and cease to 'tick'. Adding the probable number of dead pulsars to those that we can now detect brings the number up to a level much greater than that of probable supernovæ.

If our Galaxy contains pulsars, then presumably other galaxies do so also, and this has been confirmed very recently from the Parkes observatory in New South Wales, where Dr John Ables has identified a pulsar in the Large Cloud of Magellan. This is of special importance, because the way in which the ticking signals are affected by their long journey from the Cloud to ourselves – some 180,000 light-years – tells us a great deal about the nature of intergalactic material. No doubt pulsars will be found eventually in more remote systems, such as the Andromeda Spiral, though it will mean using equipment more sensitive than anything we have yet managed to build.

I have said a good deal about pulsars, and galactic radio sources in general, because in modern astronomy they are of tremendous importance, but it is a fact that the greatest numbers of identified radio sources lie beyond the Milky Way. I have referred to the exploding galaxy Messier 82. Another strong radio source is the giant elliptical galaxy Messier 87, in Virgo, from which issues a curious jet of material. The list by now is very extensive indeed, and this leads on to perhaps the most startling objects of all, the QSOs or quasars. But it is fitting to end the present chapter with an event which occurred in the autumn of 1981: Sir Bernard Lovell retired as Director of Jodrell Bank. It marked the end of an era, because Sir Bernard has been the greatest of the early pioneers. I believe that when the history of 20th-century science comes to be written, he will be regarded as the Isaac Newton of radio astronomy.

# 16 The Quasar Story

Astronomers, as a class, tend to become used to surprises. There is always something new 'just around the corner', and almost every type of celestial object produces its quota of the unexpected, the ticking pulsars and Saturn's grooved rings being good examples. But the biggest shock of all, I think, must have come in 1963 to Dr Maarten Schmidt and his colleagues at Palomar, when they suddenly found themselves face to face with objects that were not only unexpected, but absolutely incredible: QSOs or quasars.

Way back in 1960, at a meeting of the American Astronomical Society, Dr Allan Sandage reported that a strong radio source, 3C-48 (so called because it was the 48th object in the third Cambridge catalogue of radio sources) had been identified with a very faint star in the constellation of the Triangle. Associated with 3C-48 was what looked like a wisp of gas. Sandage said that he had looked at the spectrum and found it to be unlike anything else he knew; perhaps 3C-48 might be an unusual supernova remnant, or even a distant galaxy? Two more sources of the same kind were found in the following year, and several more in 1962. Astronomers in general were not particularly excited, and the most popular theory was that the objects really did represent supernova remnants, but they were of some interest, and were worthy of more than casual attention.

The main trouble about the radio astronomy of 1963 was, as we have seen, the lack of resolution. Not only was it impossible to separate two radio sources lying side by side, but it was also impossible to give the position of a single source with any real accuracy, and it was not good enough to say merely that a radio emitter was 'somewhere around here'. This, in turn, meant that the identifications with stars or starlike objects was uncertain. Fortunately, Nature came to the rescue. One of the new sources, 3C-273 (incidentally, the brightest member of the class; it had been known for many years and had been listed as a normal star) was so placed in the sky that it could at times be covered or occulted by the Moon. This would give a golden opportunity for measuring its precise position. As the Moon swept across the source, the radio waves would be cut off, and since the position of the Moon was known the position of the source would automatically follow.

Nature co-operated only in a rather reluctant way. The forthcoming occultation of 3C-273 would not be visible from England, where by this time Jodrell Bank was in full operation, or from America. The only observing site would be Australia. This meant calling in the radio

Making my way into the 210-foot dish of the Parkes radio telescope, photographed by Pieter Morpurgo.

The Parkes 210-foot radio telescope; my photograph, December 1981.

astronomy observatory at Parkes in New South Wales, a few hours' drive from Sydney, and within easy reach of Coonabarabran (though at that time, of course, the AAT had not even been planned, and Coonabarabran had not become world-famous).

At Parkes, a 210-foot 'dish' had been set up, second in size only to that at Jodrell Bank, and with the advantage of being able to study the southern sky. The telescope is still in full operation, and is perhaps even more useful than it was in 1963, so that it is worth pausing to say a little more about it.

As I drove up to the Observatory in late 1981, on my way to Sydney from Coonabarabran, the 210-foot dish came into view some way from the Observatory boundary. It is a truly imposing structure. At Jodrell Bank, the main offices and control rooms are in separate buildings; at Parkes they are in the tower upon which stands the dish itself. When lowered, the bowl of the dish comes within a few inches of the ground, so that you can walk into the bowl towards the central equipment – or, rather, climb up on steps; when the bowl is turned slowly upwards you have the impression of being cut off from the world, because nothing is visible except for the rim of the bowl around you and the sky above. It can also, incidentally, be very hot. The summer sun beat down while I was in the bowl, and the material covering the metal 'network' was strongly warmed.

The mounting of the dish can be used either as an equatorial or as an altazimuth; part of the design, I learned, was due to no less a person than the great inventor Barnes Wallis. One special feature is the method of guiding. In the heart of the radio telescope there is a small optical reflector with a 4-inch mirror which never sees the sky, but is aimed at a mirror on the underside of the bowl. To set the dish, the optical telescope is first pointed to the required direction, which is easy enough and was no problem with the relatively primitive computers of 20 years ago. Once this has been done, the dish moves automatically until the optical telescope 'sees' itself. Optical telescope and dish must then be pointing in exactly the same direction. So far as I know, Parkes is the only observatory to use this method, but it is remarkably effective, and was incorporated in the original plan; the whole radio telescope was completed in 1961, only four years after Jodrell Bank, and there has been no need to modify it.

There is also a smaller radio telescope mounted on rails, so that it can be moved around and used with the 210-foot to make up an interferometer. But it is the great dish that is all-important, and it made history in 1963 when it was first turned towards 3C-273.

Actually, things were not so straightforward as had been expected. The astronomers – C. Hazard, M. Mackey and A. Shimmins – knew the position of the radio source within fairly acceptable limits, but when they worked out the altitude above the horizon they found that they were faced with a real problem. 3C-373 would be so low down at the time of occultation that the dish could not be tipped far enough to reach it. Panic! What was to be done? Hazard and his colleagues decided upon drastic action. Local obstructions were ruthlessly hacked down, part of the gearing of the telescope's mechanism was dismantled, and some of the teeth were filed off. When everything possible had been done, it was estimated that the emersion from occultation would *just* be observable. It was; and for the first time, the position of a radio source was pinpointed very accurately. 3C-273 was indeed identical with a rather dim bluish star, but instead of a single source there were two, one to either side of the visible object.

The next step was to find out what the 'star' was really like. Using the Palomar 200-inch, Maarten Schmidt photographed the spectrum, and settled down to analyse it. It was then that he had what may justifiably be called the shock of a lifetime. 3C-273 was not a star at all. The spectrum was quite un-stellar, and the absorption lines due to hydrogen were tremendously red-shifted, which could only mean that the object was very remote – perhaps thousands of millions of light-years away. If so, then it must also be much more luminous than any known galaxy, not by two or three times, but by tens or hundreds. Yet it was small, which is why it had always been taken for a star. At first, it and similar sources were called QSOs or Quasi-Stellar Objects; before long this was shortened to 'quasar', though both terms are in use today, and latterly QSO seems to have come back into favour.

It is probably true to say that Schmidt's announcement caused as much interest in astronomical circles as anything since Hubble's work of 1923. Quasars appeared to be completely illogical. How could a comparatively small object shine with such amazing brilliance? The hunt was on, and more and more quasars were found, some of them much further away and

Quasar; 3C-273 in Virgo, the only quasar bright enough to be seen with relatively small telescopes. Photograph taken with the largest telescope at Kitt Peak

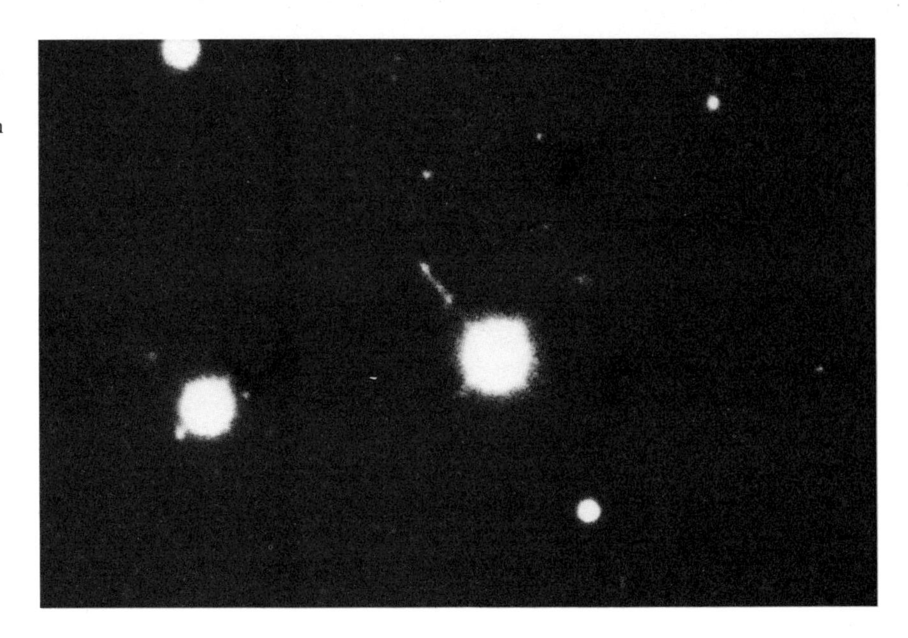

therefore presumably more luminous than 3C-273. By now over a thousand are known, and the most remote so far measured, OQ 172, is estimated to be between 15,000 million and 20,000 million light-years from us. Nearly 600 quasars have been discovered from Parkes, and the optical identifications have been made at Siding Spring with the AAT.

Originally the quasars were tracked down because of their radio waves; otherwise nobody would have paid much attention to 3C-273, and it will still have been listed as a normal star in our own Galaxy. Yet we now know that not all quasars are radio emitters. Many of them are radio quiet, though it does look as though the most remote and powerful quasars are also radio sources. X-rays and infra-red are also very much in evidence, and the Einstein Observatory results showed that some quasars are very variable in the X-range. On one occasion, for example, the quasar OX-169 showed a decrease of 30 per cent over a period of only a few hours.

Once the strange characteristics of the quasars had been established, the pressing need was to find out just how and why they shone so powerfully, and all sorts of theories were put forward. We must be honest enough to say that we still do not know the answer, and all we can do is to weed out those theories that do not fit the facts.

First, can it be that quasars are powered by successions of supernovæ? Every supernova becomes at least 15,000,000 times as luminous as the Sun, and sufficient numbers of supernovæ might add up to a quasar. However, supernovæ do not last for more than a few months, or a few years at most; there would have to be vast numbers of them in a single quasar, and there is no reason to believe that one outburst will trigger off another in a kind of chain reaction.

Secondly, there was a suggestion that the basic cause might be a sort of 'supercluster' of exceptionally luminous stars. But this is not convincing either, because the total light-emitting area would have to be much larger than a quasar.

Thirdly, quasars could be regions of space in which matter was appearing spontaneously out of nothingness; 'White Holes' in fact. This was originally proposed by Sir Fred Hoyle, and outlined by him in a television programme transmitted by the BBC in December 1964. He suggested that matter might appear in concentrated blobs, and that by some unknown process an immense mass appeared in the same area at the same moment, producing a quasar. Not many astronomers were impressed; the theory of the continuous creation of matter was coming under heavy fire even then (as we shall see later), and in any case there seemed to be so many quasars that an impossibly large amount of new material would be involved.

Next came the idea of anti-matter, assumed to be the complete opposite of ordinary matter. This was due initially to the great Swedish physicist Hannes Alfvén, and it is at least intriguing. An Alfvén universe is made up partly of ordinary matter (koinomatter) and partly of anti-matter. When the two types meet, they annihilate each other in flashes of energy. There might be whole galaxies made up of anti-matter, and to us they would look quite normal, but if two opposing regions collided there would be wholesale mutual annihilation, with both koinomatter and anti-matter being turned into radiation. In this case, a quasar might be the result of an unfriendly meeting between a koinomatter galaxy and an anti-matter galaxy. I doubt whether this theory will ever be either conclusively proved or denied, because there seems to be no possible way of setting up an experiment, but the existence of Alfvén-type anti-matter rests upon no definite evidence, so that the whole idea seems implausibly speculative.

Lastly, a quasar could be powered by a gigantic Black Hole, sucking in material from all around and swallowing up star after star. This is more promising, and is to some extent supported by the fact that so many quasars are also X-ray sources.

Black Holes had not come on the scene when quasars were discovered, but since then there has, I feel, been a tendency for some astronomers to regard them as remedies for everything. On the other hand there is a good deal of indirect evidence for Black Holes inside quasars, and in some galaxies as well – notably the Seyfert galaxies, which have condensed, bright nuclei and weak spiral arms. Messier 87, the giant elliptical in the Virgo cluster, is another candidate.

It is often believed that there is a Black Hole in the centre of our Galaxy, though investigations are not easy because the galactic nucleus lies beyond the lovely star-clouds in Sagittarius, and no visible light from it can reach us. To study the centre of the Galaxy we depend upon infra-red, radio waves and X-rays, which can slice through the intervening dust (gas is not nearly so effective as a screen). Beyond the Sagittarius star-clouds it seems that there are 'arms' and 'clouds' expanding outwards from the galactic nucleus, beyond which is a disk about four light-years across evenly illuminated by some central source which may be a Black Hole, acting as a sort of cosmic plughole, so that material swirls around it before being drawn across the event horizon and permanently cut off.

Recent work has been carried out at the Royal Greenwich Observatory

Dr Alan Wright at Parkes. I took this photograph with the smaller of the Parkes radio telescopes in the background; December 1981.

and with the AAT at Siding Spring using what is termed a CCD, or Charge-Coupled Device, developed largely by Dr Peter Young, who went from England to California and whose death in 1981 is so deeply regretted. A CCD consists of a silicon chip 10 × 15 millimetres in size, with 16,000 picture elements or 'pixels' etched on its surface (one can easily appreciate how difficult it is to make). Light falling on to the chip is converted to an electrical charge, and fed into a television screen via a computer. The device is amazingly sensitive to infra-red, and has been used to show that there is a double source at the galactic nucleus; the two components are separated by a third of a light-year, and it has been proposed that one is a Black Hole while the other is a super-star with 10,000,000 times the mass of the Sun, so that this star is losing mass to the Black Hole with the usual resulting radio and X-ray emission. If such a picture is right (and as yet there is no proof), we are dealing with what may be termed a mini-quasar.

OPPOSITE: NGC 6589-90; dust and gas in the constellation of Sagittarius, in the direction of the galactic centre. Our view of the centre of the Galaxy is in fact obscured by the huge clouds of interstellar dust between it and ourselves. The patchy nature of this obscuration can be seen from the uneven distribution of background stars across this picture. Light from the bright stars inside the dust produces the two blue reflection nebulæ NGC 6589 and 6590. while a large hydrogen cloud, mixed with some dust, glows with a characteristic red colour over most of the field of view. Photograph by David Malin, AAT

This is all very well, but so far as quasars in general are concerned one vital point must be settled: Are they really as remote and luminous as most people think?

Remember, we depend almost entirely on the red shifts of the lines in their spectra. These are interpreted as Doppler Effects, and are termed 'cosmological red shifts'. However, we must ask ourselves whether there can be any other explanation. If the red shifts are not cosmological, then quasars must be much closer to us. They are certainly outside our

Galaxy, but if they lie at distances of only a few millions of light-years we can bring their luminosities down to an acceptable level.

This is a minority view, but it is backed by some extremely eminent astronomers, including Dr Halton Arp at Mount Wilson, Sir Fred Hoyle in England, and Professor Geoffrey Burbidge, Director of the Kitt Peak Observatory. Another doubter is Dr Alan Wright, at Parkes. If this view proves to be correct, we must do some radical rethinking, and when I talked to Dr Wright recently he made some very telling comments.

Halton Arp has taken photographs that show galaxies and quasars lined up. It is rational to assume that they are genuinely associated, and should therefore be at much the same distance from us, but in some cases the red shifts are different, so either there is no genuine association or else the red shifts in the quasar spectra are misleading. Arp is one of the world's most skilful and most experienced observers, so that his techniques cannot be faulted and his opinions carry much weight, but it is fair to say that many more observations will be needed before we can decide whether or not the quasar-galaxy alignments can be put down to coincidence.

Several instances of quasar pairs, in which the two objects appear to be perfect twins lying side by side, are known. To account for this, it is generally believed that what we are really seeing is a double image of the same quasar, in which case the light from the remote single quasar has been split up by something lying between it and ourselves – possibly a Black Hole. We are therefore seeing a sort of gravitational lens effect, which is plausible enough.

Alan Wright, who has been using the Parkes 210-foot dish for his researches, told me that on the whole he believed in the cosmological theory of the quasar red shifts on Monday, Tuesday and Wednesday, but not on Thursday or Friday, because there is considerable evidence in support of both theories! One significant factor concerns our old friend 3C-273, which is the only quasar bright enough to be seen with a small telescope, though of course it looks exactly like a star. Recent work in the United States has shown that the two radio sources in it are racing apart so quickly that if the quasar really is as remote as the red shifts indicate, the individual sources must be moving out from it at ten times the velocity of light, which may safely be ruled out because it would overturn the whole of existing physics. Therefore if these measurements are correct, 3C-273 must be relatively local to our Galaxy.

Another significant factor concerns 'the remotest object known'. The present official holder of the record is a quasar known as OQ 172. On the cosmological theory it must be speeding away from us at around 95 per cent of the velocity of light, and must be close to the edge of the observable universe, at least 15,000 million light-years away. OQ 172 was discovered several years ago, and is by no means the faintest known quasar, and yet no objects at greater distances have been discovered, either by Wright or by anyone else. It is almost as though with OQ 172 we have reached the actual limit, so that there are no quasars beyond, but this raises obvious difficulties. If OQ 172 were two or three times as remote as its red shift indicates, we would still be able to see it. Why, then, have we found none at still greater distances? Are they disguised in some way? Are they really absent? Or – are our distance-estimates grossly in error? It has even been suggested that OQ 172 is as close as the well-known radio source Centaurus A, in which case it is, cosmically speaking, a near neighbour. Finally, Wright draws attention to the fact that whereas the closer quasars may be either radio noisy or radio quiet, all those with high red shifts are strong radio emitters, so that presumably they are more active and more powerful.

There is a growing belief that quasars are really nothing more nor less than the nuclei of very active galaxies. This does not make their energy-sources any the less puzzling, but it does indicate that quasars, active galaxies such as Seyferts, and more normal galaxies may be three stages of evolution of the same class of object, as I remember Sir Martin Ryle saying at a meeting of the International Astronomical Union in Prague back in 1966. The 'galaxy' parts of a quasar are bound to be extremely faint and overpowered by the bright centre, but modern techniques have been able to detect them, which is highly significant.

We must also consider the strange BL Lacertæ objects, which may be intermediate in type between quasars and active galaxies. Again it was Maarten Schmidt who took the first step, when he realized that one particular radio source could be identified with what had always been thought to be an ordinary variable star in our Galaxy, BL Lacertæ in the little constellation of the Lizard. When he looked at the optical spectrum, he saw absolutely nothing except a rainbow band. There were no lines across it, either bright or dark, and therefore there was no way of estimating its distance. Further efforts met with an equal lack of success, and, as David Allen commented wryly, BL Lacertæ had won the first

BL Lacertæ, prototype of this strange class of objects.

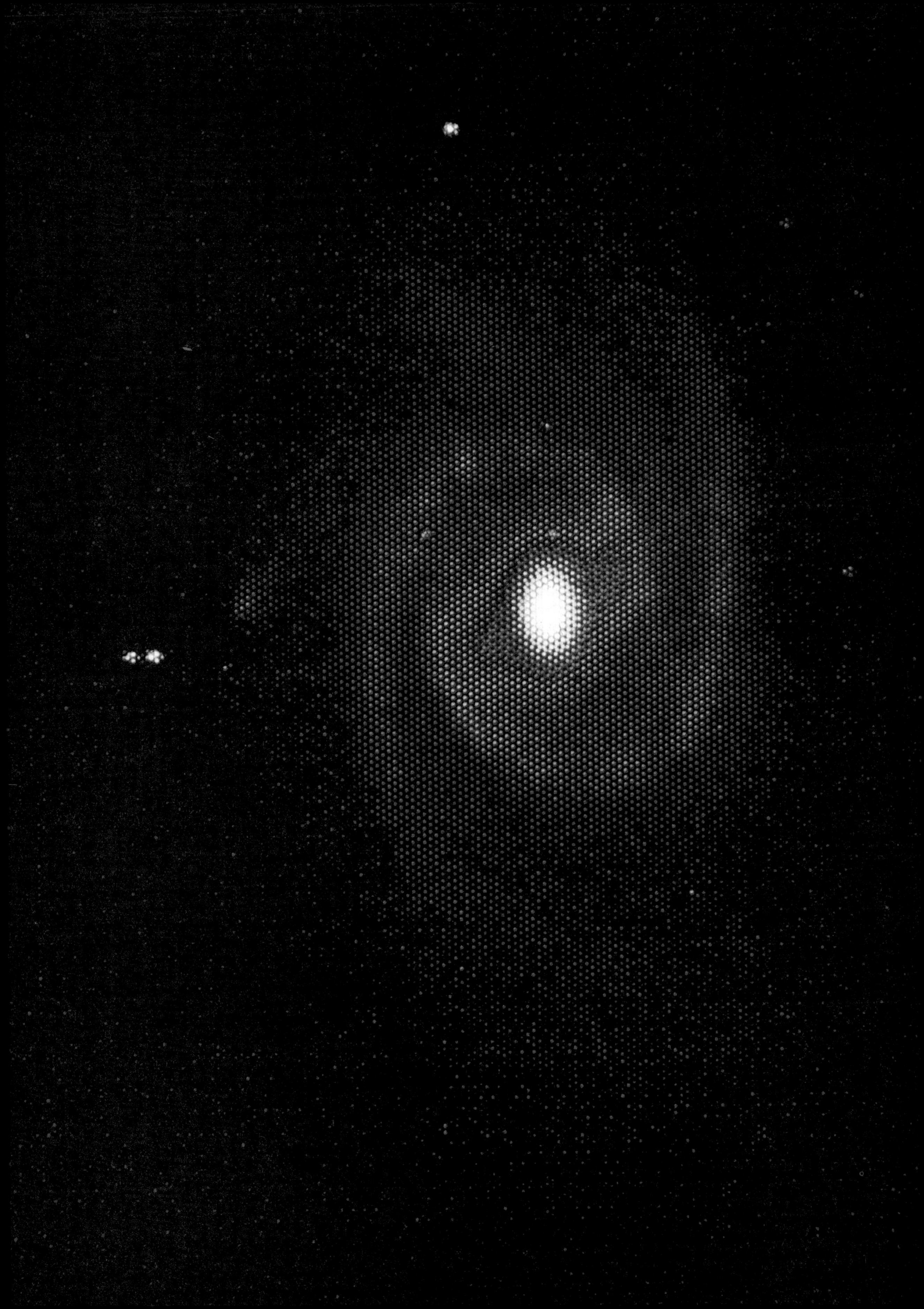

NGC 4156, a beautiful barred spiral galaxy. It has a visual magnitude of 14, and since it is very remote it must also be exceptionally luminous for a spiral of this kind. Photograph taken by P. Waddell at Kitt Peak, Arizona.

round. Subsequently other BL Lacertæ objects were found, though they were much less common than quasars.

The problem was resolved by J. B. Oke and J. Gunn in 1974, when they adopted a new line of attack. They 'blacked out' BL Lacertæ itself, and then took photographic spectra of the region of sky close beside it. This time a few spectral lines were detected, including the famous H and K lines due to calcium, and from the red shifts they estimated that the distance was about 1200 million light-years. BL Lacertæ, then, was the heart of a very dim galaxy, and this was also found to be the case for other members of the class.

Alan Wright believes that a BL Lacertæ object is simply the 'naked engine' of a quasar, and shows a different type of spectrum only because there is much less dust and gas around it. Yet he is doubtful whether quasars and BL Lacertæ objects are true relations of normal galaxies. As he has commented, a firecracker and a nuclear bomb both go off bang, but that does not mean that they are identical in nature.

Whatever may be the truth of it, we have to agree that we still do not know why these remarkable objects are so luminous. The Black Hole theory is probably the best to date, but it has yet to be proved. It is rather ironic that though quasars were identified several years before pulsars, we still know much less about them.

Summing up, it is believed by most astronomers that we really are dealing with super-luminous, intensely remote objects, some of which lie not far from the edge of that part of the universe which is open to our inspection; but definite doubts remain, and at any moment we may have to modify our theories drastically.

Even if the red shifts are not cosmological, quasars are still a long way away. It is notable that no quasar has ever been found to show a blue shift, and the idea of an expanding universe seems to be so firmly rooted that we may regard it as at least 99 per cent definite. In what follows I am going to assume that the majority opinion is right, and using all the methods available to us – optical astronomy, radio astronomy, infra-red, short-wave and mathematical analysis – we must see whether we are yet in a position to discuss the most fundamental problem of all: How did the universe begin?

# 17 The Expanding Universe

When George Ellery Hale planned the great Californian reflectors, his constant call was for 'More light!' because he knew that only by studying excessively faint objects can we look out into the depths of the universe. Since Hale's time we have been able to use methods which to him were quite unknown, and we have pushed our investigations out to more than 10,000 million light-years. What we want to know next is: Can we go on probing indefinitely, or will we come to a definite and impassable limit?

Let us come back to Hubble's Constant, according to which the recessional velocity of a galaxy or quasar increases according to its distance from us. (I repeat that I am following the majority opinion that the spectral red shifts really are cosmological.) If the remoter galaxies run away from us more and more quickly, we will eventually come to a distance at which they are receding with the full velocity of light, so that we will not be able to see them; we will have come to the edge of the observable universe, though not necessarily to the edge of the universe itself. Of course, this does not mean that we are in any special position. Every group of galaxies is receding from every other group.

Looking through contemporary literature, we find different estimates of the critical distance at which a galaxy or quasar will be receding at the velocity of light. It is certainly over 10,000 million light-years. A favoured value is 15,000 million light-years, though some astronomers increase this to 20,000 million. For the moment I propose to be conservative, and take a mean value of 15,000 million light-years, with the proviso that this may be either slightly too large or considerably too small.

NGC 4676 A and B, known as the 'Mice' galaxies because streamers of stellar matter distort their shapes. These interacting galaxies were photographed with the 158-inch Kitt Peak reflector and then computer-enhanced for detailed study; colour coding, added by Kitt Peak's Interactive Picture Processing System, shows relative brightness, with the greatest intensity at the centre and the least intensity at the outer edges of the two galaxies. The bridge of matter connecting the pair is believed to be hydrogen.

When you remember that a light-year is equivalent to almost six million million miles, it is evident that the size of the observable universe is quite beyond our understanding. But even this does not mean that we have the full picture. Presumably an observer close to a remote quasar would be able to 'see' a region well beyond our own limit, to the opposite side; and whether or not the universe is finite is a problem which is very difficult to answer.

Actually, we are in trouble whichever attitude we adopt. If the universe is finite, then what is outside it? Answering 'Well – nothing', is simply a cowardly evasion, because in this context 'nothing' is the same thing as 'space'. Yet if the universe is not finite, we are faced with the necessity of imagining something that goes on for ever, which is impossibly mind-boggling. There are mathematical models which assume space to be finite but unbounded, but putting this concept into plain

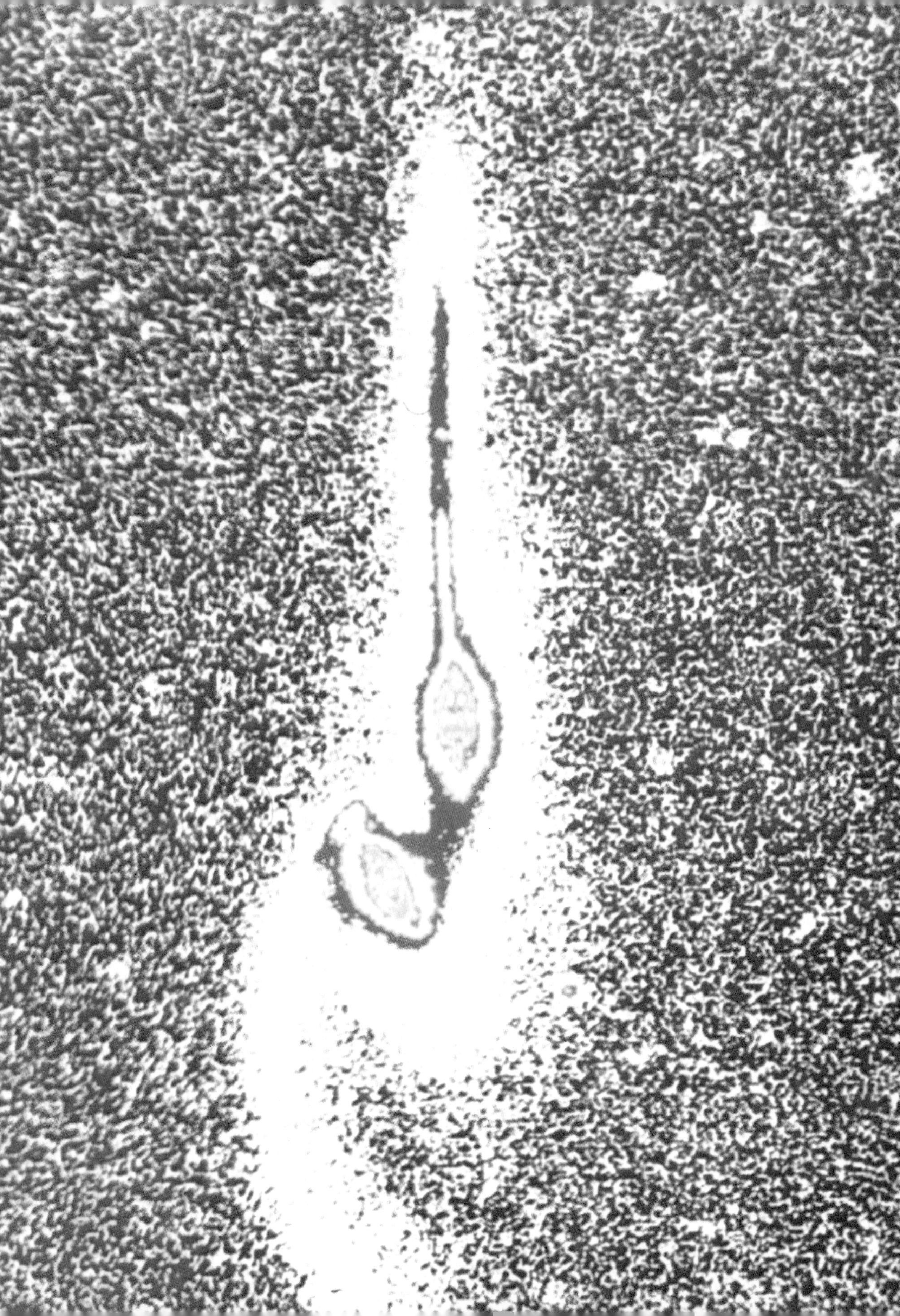

The Coma cluster of galaxies; Cerro Tololo photograph. The cluster contains many galaxies of various types. Its distance from us is of the order of 100 megaparsecs – one parsec being equal to 32.6 light-years and one megaparsec being equal to a million parsecs. It is hardly surprising that most of the galaxies in the cluster appear as nothing more than tiny blurs of light, even though some of them must contain at least 100 thousand million stars.

English is, obviously, hard. On one of the few occasions when I met Albert Einstein I asked him whether he could give me a non-mathematical, really coherent explanation of finite but unbounded space. He replied that he could not; and where Einstein failed, I am hardly likely to succeed.

We are in the same trouble when considering the nature of time, but at least we have some idea of the scale involved, because there are certain reliable guides. All the evidence indicates that the age of the Earth is between 4500 and 5000 million years, and that of the Sun rather over 5000 million years. Our Galaxy must therefore be at least 10,000 million years old, and we may assume that it dates back from 15,000 million to 20,000 million years. Working backwards, so to speak, we can calculate that all the galaxies were crowded together between 15,000 million and 20,000 million years in the past. Current ideas indicate that all the material in the universe was created at one instant, in the form of what may be termed a primeval atom. There was a violent explosion – which is why we refer to the Big Bang theory – and expansion began. Heavier elements were built up from the original hydrogen; galaxies evolved; stars condensed out; planetary systems were formed; life began; and the result was the universe as we know it today.

During the 1950s this picture was challenged by a totally different theory, due originally to Hermann Bondi and Thomas Gold, but elaborated and extended by Sir Fred Hoyle. This time there was no big bang, and for that matter no beginning. According to the so-called continuous creation or steady-state theory, the universe has always existed, and will exist for ever. As old galaxies pass over the boundary of the observable universe new ones take their place, produced by the spontaneous creation of new hydrogen atoms out of nothingness. The rate of creation would be too slow to be measurable, but the theory could be checked by observation, because it assumed that the universe would always be in a steady state. If we could come back in, say, a million million years' time we would still see the same number of galaxies as we do today, even though they would not be the same galaxies.

To check this properly would mean travelling backwards in time, and in my opinion this is one of the few things which is really impossible. Fortunately we can do the next best thing, and look backwards in time, because when we see a galaxy or a quasar at a distance of, say, 10,000 million light-years, we are seeing it as it used to be 10,000 million years ago. By examining very remote objects, we can see the universe not as it is now, but as it used to be in the remote past. On Hoyle's theory the galaxies would still be spread around in the same way. But this is not so; there are definite differences in distribution, so that the universe is not in a steady state, and the whole theory is wrong.

If we revert to the big bang idea, it is obvious that immediately after the creation of matter in the primeval atom the overall temperature must have been very high, so that the youthful universe was a very hot place indeed. In 1965 Dr Robert Dicke and his colleagues in America were wondering whether it might be possible to detect the 'left-over' heat from this initial outburst, and were working on equipment to search for it, when A. Penzias and R. Wilson independently announced that they had detected radiations at a wavelength of 3.2 centimetres, coming from all directions in space. This would indicate a general temperature of 3 degrees above absolute zero (absolute zero being the coldest state possible), and the results fitted in so well with Dicke's theory that it seems as though we really are measuring the remnant of the big bang.

Since then some new information has come in, as a result of co-operative studies made at two great observatories: Palomar in California, and Siding Spring in Australia. The equipment used is known officially as an IPCS or Image Photon Counting System, though since it was developed in England by Professor Alec Boksenberg, now Director of the Royal Greenwich Observatory at Herstmonceux, it is more generally called the Boksenberg Detector. It involves a special intensifying television camera that amplifies the extremely faint images collected by the telescope by more than ten million times, so that the individual light-photons can be counted. (Remember that for some purposes it is convenient to regard light as a stream of small parcels or photons rather than a wave-motion, though one must never fall into the trap of picturing the photons as solid lumps.) When used together with a large telescope, the power of the Boksenberg Detector is truly awesome.

Boksenberg and his colleagues in America, Professor Wallace Sargent and the late Dr Peter Young, used detectors on the Palomar 200-inch reflector and the AAT at Siding Spring to look for diffuse clouds of hydrogen lying between the galaxies, so representing pristine material left over from the big bang. They looked first at quasars, the most remote objects known. As we have seen, the spectral lines of quasars are tremendously red-shifted, but the investigators found that there were also absorption lines due to hydrogen which did not share in this shift. Therefore the hydrogen lines were due not to the quasars themselves, but to hydrogen lying between the quasars and the Earth.

Absorption lines due to carbon were absent, which proved that the clouds were extremely ancient. As supernovæ explode, they hurl their material away into space, and contaminate the original hydrogen with carbon and other elements. With Boksenberg's intergalactic clouds there was no carbon, and so no contamination; they were indeed left over from the big bang that had occurred around 15,000 million years ago, instead of having been ejected from supernovæ or quasars, as had been previously believed. They were floating independently in intergalactic space. Inside galaxies, the material between the stars is polluted by heavier elements instead of consisting only of hydrogen. Outside the galaxies, it remains the same as it used to be in the first period of the existence of the universe.

All in all, it seems that there must certainly have been a sudden creation between 15,000 and 20,000 million years ago, followed by rapid expansion of the material which had so suddenly come into being. It is pointless to ask just where in space this event took place; space was created at the same time as the material, which means that the big bang happened 'everywhere'! But can we be sure that even this event marked the actual beginning of the universe?

We know that the galaxies are moving outwards, and that their speeds increase with distance. What we cannot yet decide is whether the Hubble relationship holds good to the extreme limit. In other words, do galaxies start to slow down in the remote parts of the observable universe, finally pausing before turning back once more? If so, then the present phase of expansion will be followed by one of contraction. The galaxies will rush inwards, so that after perhaps 80,000 million years they will meet up and there will be a new big bang, followed by another phase of expansion. This gives us a loophole for supposing that the universe has existed for an infinite time in the past.

Here there is a possible check, because everything depends upon the mean density of the universe. If there is enough material, the receding galaxies will not be able to achieve full escape velocity, and eventually they will turn round and come back. If the material is insufficient, the expansion will continue indefinitely, until at last all communications between the groups of galaxies will be severed.

The critical density is surprisingly low, and amounts to a mean of one atom of hydrogen in a volume twice as great as that of the Earth's globe. Preliminary calculations show that there is more material than this, ruling out the oscillating picture. But wait! I have earlier referred to the 'missing mass' in clusters of galaxies; there seems to be more material than we can actually detect, and it could be locked up in Black Holes, though there are other possibilities as well. For instance, there could be unexpectedly large numbers of dwarf stars too dim for us to see; it is also possible that the elusive neutrinos have slight mass, which in view of their quantities would make a great deal of difference.

The Toadstool Galaxies, ESO B138 and IG 29 and 30. This photograph was taken with the Kitt Peak 158-inch reflector and then computer-enhanced; note the bridge of hydrogen connecting the two large galaxies, inside which star formation is believed to be in progress.

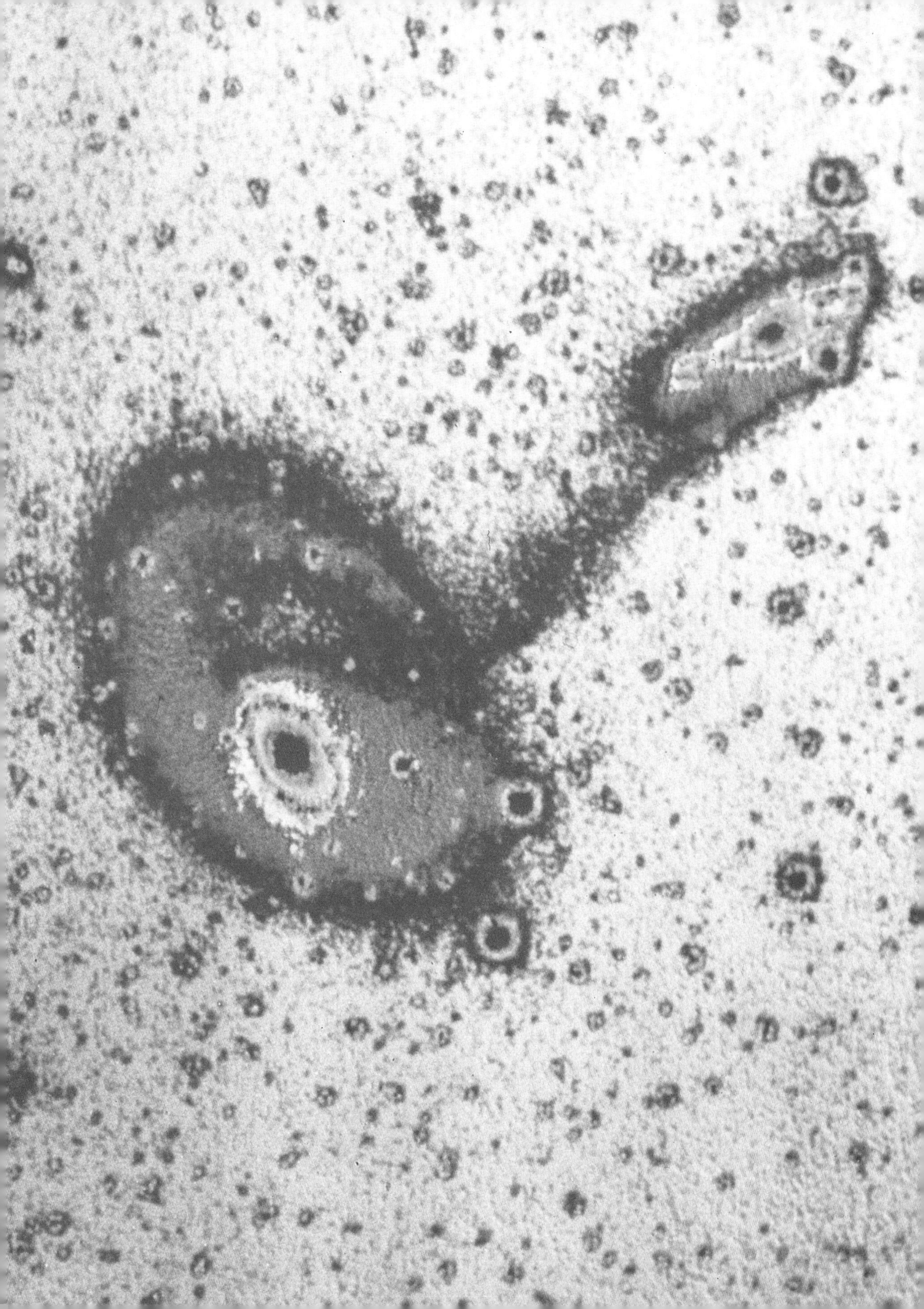

At the moment, the consensus of opinion is that the overall density is too low, that the galaxies will never approach each other again, and that the universe will evolve to a situation when all the clusters of galaxies are out of touch with each other. Their member stars will die, and we will be left with dead worlds, lightless, heatless and silent. It is not a pleasant picture; but the whole question is still open, and very recent theoretical work by Dr John Beckman and Mark Kidger at Queen Mary College, University of London, suggests that the density may after all be just enough to prevent indefinite expansion.

At least we may be sure that long before the universe reaches this depressing state (if it ever does), life here, and even the Earth itself, will have been destroyed. When the Sun leaves the Main Sequence and becomes a Red Giant, as it must do in 5000 or 6000 million years time, there can be no hope for us. Whether this means the end of all intelligent life in the universe is quite another matter, and here too opinions have changed markedly of late.

We have learned a great deal about living matter, but it would be conceited to claim that we know much about the nature of life itself, and we cannot create it (though, unfortunately, we can destroy it with alarming ease, as two world wars over the past 70 years have shown). Sir Fred Hoyle and his colleagues believe that life originated far away in space, and was brought here by means of a comet or some such agency. Most other people prefer to believe that life on Earth began on Earth. Either way, the creation of living matter from non-living matter requires

NGC 6520 and the dark nebula Barnard 86. This photograph shows varied regions in space, with many stars and also a mass of nebulosity which is dark because there are no suitable stars within it to make it shine. Sir Fred Hoyle maintains that life began well away from the Earth, and it is certainly true that many organic molecules have now been detected in interstellar space.

a whole series of improbable coincidences. This being so, are we likely to find life anywhere else in the universe?

Let me stress at once that I am discussing only life as we know it. Matter is made up of atoms and molecules, and the only atoms that can link up to produce living matter are those of carbon. (Silicon atoms can make some sort of attempt, but as yet there is no evidence of silicon-based life.) We know that all the bodies we can see in the universe, out to the remotest galaxies and quasars, consist of familiar elements such as hydrogen, oxygen, calcium, iron and the rest; and we can safely rule out the existence of other suitable atom-types (if not, the whole of our chemistry and physics is wrong). It follows that life, wherever it is found, will be based on carbon, so that it will need the right kind of temperature, the right kind of atmosphere, suitable gravity and illumination conditions, and an adequate supply of water. As soon as we deviate from this principle we enter the fascinating realm of science fiction, with alien creatures of the sort known popularly as BEMs or Bug-Eyed Monsters. Frankly, I would like to believe in BEMs; but if they exist, then we must reject the whole of modern science. When one has an incomplete set of facts, the only sensible course is to interpret them as logically as possible, and unless any contrary evidence comes to hand I, for one, will continue to believe that BEMs cannot exist on the Moon, Mars or anywhere else.

In searching for life we need spend little time on the planets of our Solar System, because most of them are so eminently unsuitable. Only Mars and Titan seem to be possible candidates. The Viking probes have shown no trace of living organisms on Mars, and Titan seems to be in a state of deep-freeze. Even if life exists on either (which I very much doubt) it is bound to be very primitive. It seems strange now to reflect that as recently as 1901 a prize was offered in France to be given to the first person to communicate with beings on another world – Mars being specifically excluded, because calling up the Martians was regarded as being too easy!

Our next step, then, is to decide whether other planetary systems exist. Unfortunately we cannot hope to see even a large planet moving round even a nearby star, but there are a few stars that seem to 'wobble' very slowly and slightly as they move through space, and these effects have been attributed to attendant planets; thus Barnard's Star, a faint Red Dwarf six light-years away, may have at least two planets, each as large as Jupiter. But here too the evidence is uncertain, and we are more or less reduced to speculation.

The only observational evidence we have of a planetary system which may be forming comes from the work of a team headed by Dr Rodger Thompson of the Steward Observatory in America. Using the Observatory's powerful reflector, and also a 36-inch reflector flown in the Kuiper Airborne Observatory, they have made a careful study of a remarkable disk-shaped, highly-luminous object known as MWC 349, in the constellation of Cygnus. It seems that here we have a star only about 1000 years old, a true stellar infant. Its surrounding disk of glowing gas is about 20 times the size of the central star, and emits about 10 times more light, while at its outer edge the disk is about as thick as the star's diameter, estimated as 90,000,000 miles. The star itself has 30 times the mass of the Sun, and is expected to shine for only about 100 million years, because it is squandering its nuclear fuel at so rapid a rate. The significance of all this lies in the fact that planets may even now be forming in the luminous disk, or just outside it. The distance from Earth is of the order of 10,000 light-years, so that we are now watching events which took place at about the end of the last Ice Age.

According to one interpretation, the luminous disk of MWC 349 is the inner part of a surrounding larger disk of non-luminous material, inside which planets may already have been born. The shining section is wedge-shaped, and joins the glowing surface of the parent star.

MWC 349 is massive and short-lived, but can the same sort of process have happened with our own Sun, in a greatly slowed-down form? It is at least possible. The Earth and the other planets were certainly not torn off the Sun by the action of a passing star, as used to be believed, and this is an important conclusion, because if it had been correct it would have shown that planetary systems must be very rare. The stars are so widely spread-out that there are few 'close encounters'.

The Sun is an ordinary star; what can happen to it can presumably happen to other stars also, and on the solar-nebula formation theories of today there is no reason to doubt that planetary systems are common enough. There must be plenty of Earths circling other stars, even if we cannot see them. What we do not yet know is whether life will automatically appear wherever conditions are suitable for it. In a way Mars provided a test; had any living organisms been found there we would have had a vital piece of supporting evidence, but as things are we are still in the guessing stage.

Of course, final proof would come if we could get in touch with any alien civilization. To send a rocket out to a star would take centuries at best, and so the only hope – a faint one, at that – seems to be radio contact. Radio waves, travelling at the speed of light, can flash from Earth to Alpha Centauri in little over four years, but unfortunately Alpha Centauri is not a suitable candidate; it is a binary, together with a third fainter member of the system, Proxima, which is actually the nearest known star beyond the Sun. To find a solar-type star we must go out to around 11 light-years, and turn our attention to two stars known as Tau Ceti and Epsilon Eridani. Both are visible with the naked eye, even though both are rather smaller and cooler than the Sun.

The first serious attempt at contact was made in 1960 by a team of radio astronomers at Green Bank in West Virginia, led by Dr Frank Drake. Using the 84-foot dish there, they listened out for signals at a wavelength of 21.1 centimetres, aiming the telescope at Tau Ceti and Epsilon Eridani. The choice of 21.1 centimetres was dictated by the fact that this is the wavelength of the radiation sent out by the clouds of cold hydrogen in the Galaxy, so that radio astronomers anywhere will pay particular attention to it. Not surprisingly, the results were negative, and the programme – known officially as Ozma, but more popularly as Project Little Green Men – was ended after a few months, but sporadic subsequent attempts have been made since both in America and in the Soviet Union. A more elaborate scheme proposed in the United States, involving large numbers of radio telescopes, was abandoned, largely on the insistence of

The Andromeda Spiral, M31, photographed with the Palomar 200-inch reflector. This is the nearest of the really large galaxies, at a distance of 2,200,000 light-years; it contains more than our Galaxy's total of 100,000 million stars. It is interesting to speculate as to how many inhabited worlds it contains!

a senator named Proxmire, who is strongly opposed to space research (it was rather unkindly suggested that had he lived in earlier times, he would have been similarly opposed to the development of the wheel).

Obviously there would be no chance of doing more than pick up signals rhythmical enough to be recognized as artificial, but even this would be a major step forward. It is the longest of long shots, though it is significant that it was regarded as promising enough to occupy some of the observing time of a large radio telescope.

There have been occasional sensations. One made headlines on 13 April 1963, when one British daily newspaper carried the story: it began 'Super-Civilization Discovered. Astounding Claim by Russian Astronomers.' According to this report, Dr Nikolai Kardashev and Dr Gennady Sholomitsky, in Moscow, had been studying a radio source catalogued as CTA 102, and had concluded that it was transmitting artificial signals. I heard of the report on the evening of 12 April, and in a BBC broadcast on the following morning I poured cold water on the whole idea; it was not long before the official Soviet news agency, Tass, started to play it down. It was subsequently found that what Kardashev and Sholomitsky had observed was nothing more nor less than a quasar.

Then, of course, there are flying saucers, now given the more dignified title of UFOs or Unidentified Flying Objects. They are still with us, and we continue to receive reports of cigar-shaped spaceships, flashing lights, and even alien beings who land briefly and fraternize with a chosen few before returning to their saucers. (Policemen, for some reason or other, seem particularly prone to UFOlogy.) I can only say that if I ever meet a denizen of Sirius D or Polaris K, I will make sure that they linger for long enough to join me in a 'Sky at Night' television programme!

The popular view today, among astronomers as well as laymen, is that since there is no reason to regard the Earth as unique, life is likely to be widespread. If so, our Galaxy alone must contain thousands or, more probably, millions of civilizations, some in a Stone Age, some as advanced as ourselves, and some far more highly developed and intelligent. The lack of contact can be explained by the great distances involved, unless we have been observed closely enough to be rejected as undesirable! Yet very recently the pendulum of opinion has started to swing back once more, and there is a growing view that life may, after all, be a very rare phenomenon. To this school of thought belongs Sir Bernard Lovell. Without maintaining that we are the only intelligent beings in the universe (assuming that we can regard ourselves as intelligent), he does believe that we must be very exceptional. Others go further, and claim that there is no life anywhere except on the Earth.

Whether this question will ever be answered I do not know. In the long run, it probably depends upon events here over the next century or two. If we refrain from wiping ourselves out, we may learn enough to master the secret of interstellar travel. If not, there is always the chance that in the far future a visitor from space will arrive here to find nothing more than a ruined, desolate, radioactive planet, from which all traces of life have been extinguished.

Finally, I come back to a point of view which I have given many times in the past, and which has even been regarded as a particular hobby-horse of mine. I maintain that nobody has yet discussed the origin of the universe. We must start with something; matter must have been created in some way, and no theory, steady-state, big bang or oscillating universe, has given the slightest inkling of how this happened. What we are discussing is the evolution of the universe, which is a very different thing. When it comes down to fundamentals, we are no wiser than we were 25 years ago, or indeed 25 centuries ago, and this is one problem which mankind may never be able to solve.

# 18 The Next Twenty-Five Years

Throughout this book I have been comparing our knowledge today with that of 1957. It has been a momentous period, and I feel privileged to have lived through it and to have been able to comment upon it, but looking back over the 'Sky at Night' programmes I find that some of my forecasts were very wide of the mark. Thus in 1958 I predicted that the first manned flight to the Moon would not take place for at least 15 years, and probably rather longer; actually it took less than 11. Like everyone else I was completely wrong about Mars, though I modestly claim that I was fairly accurate with regard to the Moon. Quasars, pulsars, scintars and many other types of objects had not even been considered; radio astronomy was, by modern standards, primitive, while X-ray astronomy had not even begun.

So many discoveries were made during this quarter-century that I

A telescope of the future . . . the William Herschel Telescope in its dome, due to be set up at La Palma. This is a model, since at the time of writing (February 1982), the telescope is still in the planning stage. It is quite likely that in the foreseeable future it will be operated by 'remote control' from the headquarters of the Royal Greenwich Observatory at Herstmonceux.

believe the next 25 years may prove to be in the nature of a period of consolidation, while we build on the foundations which have been laid. No doubt electronic devices will be improved beyond all recognition, just as today the Boksenberg Detector surpasses the old photographic plate. There will also be new powerful telescopes, though I rather question whether we will see any single-mirror instrument larger than the Russian 236-inch; it is quite on the cards that the multi-mirror principle, as pioneered by the MMT on Mount Hopkins, will be widely preferred.

Space research methods will obviously be all-important, and here too we have already passed the pioneer stage. The tiny artificial satellites of 1957 have given way to massive orbiting vehicles, and we have seen the first true space stations, such as the United States Skylab and the Soviet Salyuts; as I write these words (February 1982) Salyut 6 has been in orbit for more than four years, and is still fully operational. There is no doubt that the Russians are doing all they can to set up a major station which will be permanently manned, and apparently they are developing a new rocket that will be more powerful than anything we have as yet; it may be able to dispatch a 100-ton payload into orbit. The trouble is, of course, that space research and military preparations go hand in hand, and we have to admit, albeit reluctantly, that but for this the necessary funds for space projects would not have been made available, which is a sad

Skylab 4 on-board photograph. An overhead view of the Skylab space station cluster in Earth orbit as photographed from the Skylab 4 Command and Service Modules during the final fly-around before the return home. The space station is contrasted against a cloud-covered Earth. Note the solar shield, which was deployed by the second Skylab crew and which shades the Orbital Workshop in the area from which a micrometeoroid shield was missing ever since the cluster was launched on 14 May 1973. The solar panel on the left side was also lost. Inside the Command Module when this picture was taken were Astronauts Carr, Gibson and Pogue. The crew used a 70mm hand-held Hasselblad camera to take the photograph.

reflection upon human nature. It is also true that the projects now being sacrificed on purely financial grounds are also those which are the most scientifically valuable, such as the American probe to Halley's Comet.

Skylab itself did all that had been hoped of it, even though there were some anxious moments when it was launched in the spring of 1973; the first crew, made up of Astronauts Conrad, Kerwin and Weitz, were forced to carry out repairs while actually in orbit, which was in itself a major feat. Altogether three crews, each made up of three astronauts, went to Skylab, staying up respectively for 28, 59 and 84 days and undertaking an immense amount of scientific research. For example, they took over 180,000 pictures of the Sun alone. One very important aspect of the missions was that the astronauts were able to cope well with conditions of weightlessness, and suffered no more than temporary discomfort when they came down. The 84-day endurance record has long since been broken by the Soviet Salyut crews, but at the time of its active career Skylab was in a class of its own.

The only hitch came at the very end of the mission, long after the last crew had left. It had been hoped to keep Skylab in orbit until 1982 or 1983, by which time the Shuttle would have been ready, and if necessary the station could have been boosted into a high orbit, safe from the tiny but persistent drag of the Earth's upper air. Unfortunately the NASA planners had forgotten that when the Sun reaches its maximum period of activity, as it did in 1980–81, the Earth's upper atmosphere becomes appreciably thickened, and the drag is increased. As had been predicted by Desmond King-Hele, Britain's leading satellite tracker, Skylab could not survive beyond 1979, and it plunged back into the lower air, breaking up and showering fragments over Western Australia. Nobody was injured, because the fragments landed in sparsely-populated areas, but it represented a bad miscalculation.

A much worse scare had occurred rather earlier, when the unmanned Soviet satellite Cosmos 954 crash-landed in Canada. The problem was that it carried radioactive material, and there was a serious risk of contamination, which would have been much more dangerous but for the fact that the landing took place in the thinly-populated North-West Territories. It is no longer true to say that 'what goes up must come down', but the Cosmos 954 episode was a timely reminder that every possible precaution must always be taken.

As the year 1982 opens, we have to admit that in some ways the immediate outlook is rather disappointing. Ever since 1962, when Mariner 2 flew past Venus, the United States planetary missions have been in full swing, sending us back fascinating close-range information from all the members of the Solar System out as far as Saturn, with Uranus ahead. Now, with the NASA cutbacks, this programme has been if not abandoned, at least put largely into cold storage. Even the VOIR Venus orbiter, which had been scheduled for the late 1980s, has been cancelled – though two Russian probes landed on Venus in March 1982. To use an old cliché, virtually all the American space eggs have been put into one basket: the Shuttle.

One major problem of ordinary rockets is that they can be used only once. They cost millions of dollars (or roubles) to build; they make one

USA

OPPOSITE: Launch of the Space Shuttle *Columbia*, from Cape Canaveral on 12 April 1981, carrying John Young and Robert Crippen. This was the first Shuttle flight; it was completely successful.

flight into space, and that is all. No parts of them can be salvaged to be used again, which is rather like building a new Concorde for each flight between London and New York. If large orbital stations are to be established, there must be a cheaper way of going up and down to them, and the Shuttle is the best answer, because it is part-rocket, part-aircraft and part-glider. On launch, it uses two massive solid-fuel boosters, which break away and parachute down into the sea, where they can be fished out. Apart from one large external tank, the whole of the Shuttle can be used time and time again, so that in theory at least it is the perfect answer to the problem of ferrying men and supplies to and from orbit.

As so often happens, development was slow; it is almost inevitable that everything will take longer than expected (though this was not true of the Apollo programme – President Kennedy had predicted that an American would step out on to the Moon before 1970, and he was right). One trouble with the Shuttle concerned the 'tiles' fixed to the outer hull, which had to take the brunt of the heating during flight through the lower atmosphere; the outer temperature rises to over 2700 °F., so that full protection is an absolute necessity. There were other problems too, and it was not until 1981 that the first Shuttle was launched. Expertly piloted by John Young, it made a perfect unpowered landing after a successful period in orbit, and the next flight, later in the year, was also a success even though a failure in one of the fuel cells meant that the time spent in orbit was curtailed.

Whether Shuttle flights will be possible in as quick succession as had been hoped remains to be seen, but at any rate the Shuttle works, and if it had failed the whole United States manned space programme would have been held back for a very long time.

RIGHT: Touchdown of the first Shuttle flight; after a 541-hour space mission, the Shuttle lands gently on Runway 23 at the Edwards Air Force Base.

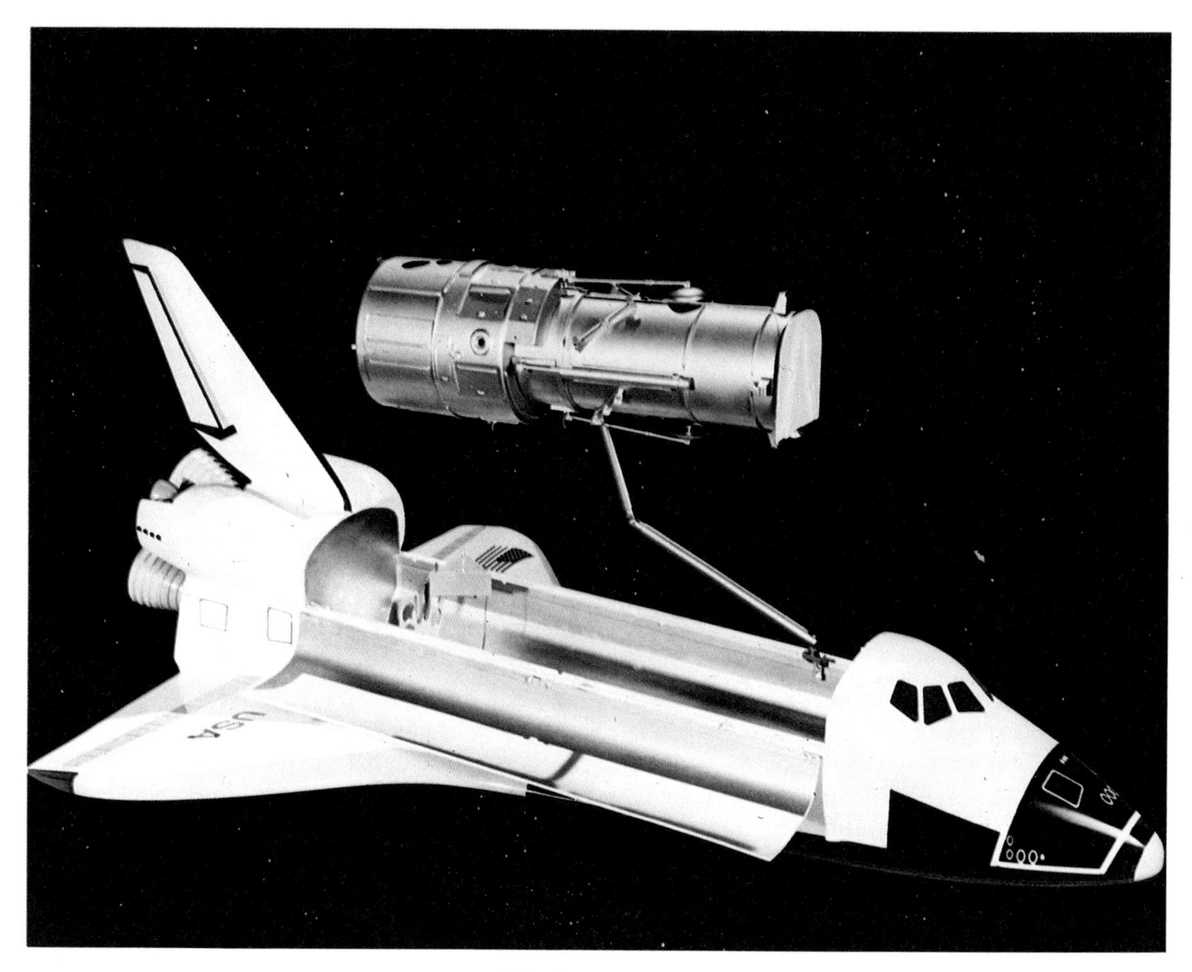


The Space Telescope; an artist's impression of the 94-inch reflector to be launched from the Shuttle.

Unquestionably the most exciting prospect from a purely astronomical point of view is the Space Telescope, scheduled to be launched in 1985. This is a really ambitious project. The telescope itself will be a reflector with a mirror 94 inches in diameter, working on a more or less conventional Cassegrain optical system. It will be taken up in a Shuttle and then separated, so that it will become an independent, free-flying satellite, moving round the world at an altitude of some 300 miles and completing one orbit every 93 minutes. Obviously it will have to be automatically controlled, transmitting its information back either directly or by way of two (or more) relay satellites which will have been put into geosynchronous orbits 23,000 miles above the equator. (This means that as the Earth spins, the satellites will keep pace with it, so remaining apparently motionless in the sky – a principle first suggested by the British scientist and author Arthur C. Clarke in 1945.)

Once the Space Telescope is in orbit it will be on its own, but the striking advances in computers and control systems indicate that there will be little real trouble in manœuvring it exactly as required so that it points in precisely the right direction. Astronomers allocated observing time with it will have the strange experience of being hundreds of miles away from their telescope instead of sitting inside it, as used to happen

30 years ago, or remaining in a control room a few yards away, as with the AAT. The current plan is that after an interval of two and a half years an astronaut will go up to carry out maintenance, and also replace any equipment that has been giving trouble or has completed its programme. After five years the telescope may be brought down for a thorough overhaul before being re-launched. How long it will last remains to be seen, but a minimum of at least 15 years is reasonable, so that the telescope should still be operating well into the 21st century.

It will carry a whole range of instruments, mainly designed for work in the ultra-violet and near infra-red part of the spectrum as well as in the visible range. There will, for instance, be what is called a Faint Object Camera, which will not be a camera in the usual sense of the term inasmuch as it will use a Boksenberg detector; there will also be spectrographs, instruments for solar research and much else.

The scope of the Space Telescope will be vast, because it will be able to record objects far beyond anything possible today, and its sensitivity should be at least 40 times that of the Palomar reflector even though the mirror will be less than half the size. It could well reach down to a magnitude of around + 27, and it may even detect planets moving round Barnard's Star and a few other comparatively nearby suns. It will peer right into the heart of Halley's Comet – assuming, of course, that it is safely in orbit by the time that Halley makes its visit to the neighbourhood of the Sun. It will track the small, dim satellites of the outer planets, and no doubt discover many more. It will show details on worlds such as the Galilean satellites of Jupiter, and it will give us our first reliable information about Pluto, the sole member of the planetary family to be neglected by NASA spacecraft. More importantly still, it will have so great a resolving power that it will be ideal for stellar research.

There is, for instance, the problem of the diameters of the stars, which are difficult to measure directly because a star looks so small. Yet when the Moon sweeps over a star, occulting it, there must be a tiny time-lag before the start of the occultation and the complete disappearance of the star, and the time taken for the Moon's limb to pass right over the star will give a clue as to the star's own diameter. The method has been pioneered on Earth by the St Andrews University team led by Professor D. W. N. Stibbs, who, in 1978 used the University's telescope to show that the orange-red star Aldebaran is between 22,000,000 and 23,000,000 miles across, but things will be much easier from above the top of the atmosphere, and experiments of this kind will certainly be made with the Space Telescope.

Consider, too, the globular clusters, whose member stars are so crowded together near the centre of the system that they merge into a single mass; the Space Telescope should be able to resolve them, and study the dwarf stars in the clusters that are hopelessly beyond our range at the present time. Cepheid variables in remote galaxies will be visible, and will provide us with a vital check on the distances of the galaxies themselves, perhaps even clearing up the problem of whether the red shift method of distance-measuring really is reliable at vast distances. And with its unparalleled light-grasp it will be able to probe out to within perhaps 0.1 per cent of the estimated boundary of the observable universe. Will quasars be found that are further away than the present holder of the distance record, OQ 172? And will the telescope penetrate as far as it is theoretically possible to go?

Many of these questions may be answered within a few months of the launch, and the astronomy of today may start to look decidedly old-fashioned. If all goes well, the Space Telescope should be responsible for as great a leap forward as the 100-inch reflector did in 1917, or the Palomar reflector in 1948.

From a purely scientific point of view the United States has taken the lead, but Europe does not intend to be left out, and the seventh or eighth Shuttle flight will, it is hoped, take up Spacelab, built by the European Space Agency. Spacelab will stay attached to the orbiting section of the Shuttle, and each mission will be limited to between seven and thirty days, but Spacelab will have the advantage of being manned, so that the telescopes and other instruments will be under direct control and can be removed on landing for further immediate use. So far as the Russians are concerned, it is difficult to make forecasts simply because not much information about their astronomical projects has been released; but they are bound to have massive space stations in orbit well before 1990, and astronomy is not likely to be neglected.

Finally, we must speculate about the possibilities of more manned expeditions to the Moon, and possibly an attempt on Mars. Here, opinions differ. I have quoted Eugene Cernan as saying that nobody will go back to the Moon without new motivation, but from a purely technical viewpoint there seems to be no obstacle to the setting-up of a permanent Lunar Base well before the end of the century, and the advantages to the whole of mankind would be immense. Moreover, observing conditions on the Earth's surface are getting worse; there is increased light pollution, and radio astronomers are becoming very worried about interference at long wavelengths. Sir Bernard Lovell has even said that unless something is done, radio astronomy will come to a virtual halt within the next few decades. The best possible place for a radio astronomy observatory would be on the far side of the Moon, where all Earth transmissions would be cut off and the site would be completely radio quiet.

At least some of these possibilities should be realized; at any rate the opportunities are there, and mankind should be able to rise to the challenge. Twenty-five years hence, in the year 2007, we may hope to know a great deal more about the universe than we do today. If not, then we will have only ourselves to blame.

I must end, as I began, on a personal note. I have watched the Space Age through its pioneer period, and I hope to go on watching it for some time yet. I have presented the 'Sky at Night' programmes every month since April 1957. By April 2007 I will have reached the advanced age of 84, but if I am still around, and still broadcasting, I have no doubt that I will find plenty to say!

# Index

Page numbers in *italics* refer to illustrations and captions.